POCKET GUIDE TO
MUSHROOMS

John C. Harris

B L O O M S B U R Y
LONDON · NEW DELHI · NEW YORK · SYDNEY

CONTENTS

INTRODUCTION

This handy guide features more than 150 of the most common mushroom species found in Britain and other parts of northern Europe. It does not cover every one of the thousands of species that may be encountered – its purpose is to give an overall introduction to the wonderful world of fungi.

All the fungi here belong to the large fungi groups of Basidiomycetes and Ascomycetes. Basidiomycetes include the typical mushroom-shaped species with gills, pores or spines on the underside of the cap, as well as bracket fungi and puffballs, and club, coral and jelly fungi. Ascomycetes species, also known as sac fungi, include morels and saddle, cup, club, finger and disc fungi, and truffles.

HOW TO USE THIS BOOK

Each species featured includes a concise description, as well as photographs illustrating typical mature examples. Where necessary, further details are given in cases where appearance differs dramatically between younger and older specimens.

Some mushrooms featured are in the same genus; in these cases a brief introduction about the genus is included in the first sentence of the first species featured under that genus.

A 'quick-view' checklist after each description sums up key features: spore print (explained below), habitat, season, size and edibility. Unless otherwise specified diameter measurements are provided for caps, and height measurements followed by width measurements for stems. The mushroom identification section opposite explains what to look out for and will help you to familiarise yourself with the anatomy of a typical 'cap and stem' mushroom.

WHAT ARE FUNGI?

The kingdom of fungi, which is separate from the animal and plant kingdoms, is estimated to comprise more than a million species worldwide, many of which are very small to microscopic. The larger species we see are in fact the fruiting bodies of the organisms themselves being hidden out of sight beneath the substrate. Each such organism is made up many tiny organic fibres called hyphae, which branch out in a web-like network of fine threads called the mycelium. After a fruiting body dies, the mycelium lives on for many years – sometimes for decades or even centuries.

Young Sulphur Tuft (*Hypholoma fasciculare*)

Fungi, like animals, get their sustenance from living or dead plants and animals, because they cannot make food using sunlight as do plants. They are a vital part of the ecosystem, especially due to their ability to break down dead organic matter and create beneficial links with plants and trees in a 'mycorrhizal' relationship. Fungi are of course well known for their culinary value, but they have also provided key ingredients in the development of medicines, in the making of bread, aiding in the fermentation process, and in many important industrial uses.

MUSHROOM IDENTIFICATION
Unfortunately there is no golden rule when it comes to identifying unfamiliar mushrooms, edible or otherwise. Deadly poisonous toadstools can share the same characteristics as many edible fungi. Of course, many genera of mushrooms have familiar and reliable traits, but it is always best to know all specific features. Some species, such as the Giant Puffball, are extremely recognisable, but they are few and far between. If you are interested in mushrooms to eat, a good rule to follow is this: IF YOU DO NOT KNOW ITS NAME OR HAVE ANY DOUBT ABOUT ITS IDENTITY, DO NOT EAT IT. This is purely for safety reasons. Enjoying wild mushrooms as food or because of a casual interest in them can be very rewarding – it is just common sense to employ a smart, non-complacent attitude when dealing with new discoveries. The following information can help overcome basic identification obstacles and covers the common mushroom-shaped species – those with a typical cap and stem.

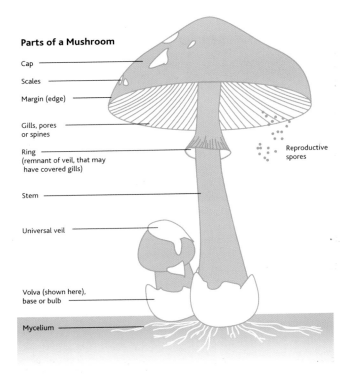

Parts of a Mushroom

Cap

Scales

Margin (edge)

Gills, pores
or spines

Ring
(remnant of veil, that may
have covered gills)

Stem

Universal veil

Volva (shown here),
base or bulb

Mycelium

Reproductive
spores

Gill attachments to the stem

Adnate: Gills widely
attached to stem

Adnexed: Gills narrowly
attached to stem

Decurrent: Gills
running down the stem

Free: Gills do
not meet the stem
(unattached)

Sinuate: Gills smoothly
notched before running
slightly down the stem

Emarginate: Gills
notched before
attaching

1. ANATOMY OF A MUSHROOM

Note that all parts shown in the diagram opposite are not necessarily those featured on all mushrooms – for example, some species may have no scales on the cap, or rings on the stem, and so on.

Useful details featured on different parts of a mushroom, which are often missed, help considerably in identification. If possible, try to make notes of all the following:

CAP

Size and Shape. What are the dimensions – the width and depth? Is it flat, round, domed or some other shape? The following illustration shows cap morphology. Note that one or more features are sometimes present on the same mushroom: for example, some funnel mushrooms also feature a bump (or umbo) at the centre. These are the different cap shapes that are commonly encountered.

Cap morphology

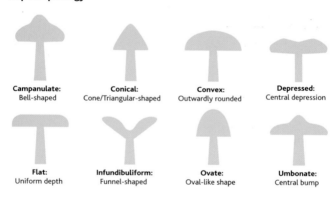

Campanulate: Bell-shaped

Conical: Cone/Triangular-shaped

Convex: Outwardly rounded

Depressed: Central depression

Flat: Uniform depth

Infundibuliform: Funnel-shaped

Ovate: Oval-like shape

Umbonate: Central bump

Colour What is the colour (or colours) of the cap? Does it change colour when bruised or handled? Cut the cap in half – does it change colour when exposed to the air?

Texture/markings Is the cap smooth, sticky, shaggy, scaly, velvety or of some other texture?

Consistency Is the cap hard, firm, fleshy, soft, spongy, fibrous, delicate, tough, crumbly or of some other consistency?

Odour What smells can you detect, if any (mushroomy, floury, chemical-like, fruity and so on)?

MARGIN (EDGE OF CAP)
What features does the margin have (if any). Is it wavy, split, inrolled (curving inwards), turned up, striated (striped, grooved or ridged)? Have remnants of the partial veil been left?

GILLS, PORES OR SPINES
The underside of a typical mushroom with a cap and stem may have gills (blade-like flesh), pores (small holes from the ends of tubes that are hidden within the cap) or even spines (hanging spikes or teeth) from which the reproductive spores will drop.

Gill attachment Gills sometimes attach themselves to the stem and sometimes not. This is an important feature to take note of. The cross-section diagram on page 8 illustrates the different ways in which gills or pores can be attached in relation to the stem. These terms are used throughout this guide.

Colour What colour are the gills? Colours can change during the different stages of growth (young, mature and old). Look around for young or very old examples for comparison. Does the gill colour change when marked with a knife or finger? Is it blotchy or speckled?

Spacing The gills may be crowded and thin, or thicker and spaced out.

Thickness/consistency Are the gills thick or thin? Are they fragile, strong, flexible or brittle?

Note any extra features.
For example:
  • Gills are forked (gill ends fork out) close to the margin.
  • Gills produce a 'milk-like' substance when handled.
  • Water droplets are trapped between the gills.
  • Gills attach to a circular 'collar' around the stem.
  • Gills are easily peeled or removed from the flesh.

Pores All of the boletes and polypores (bracket fungi) have pores instead of gills. These appear as small, round or angular holes on the underside of the cap. The spores develop and drop from small tubes that grow down from the flesh of the cap.
- Note the size and spacing.
- Note the colour and if they change colour on bruising (being touched or marked).
- Note if they appear as perfect circles, angular or produce a maze-like pattern.

Spines A relatively small group of fungi has hanging spines (also called teeth). This feature greatly narrows down identification.

STEM (STIPE, SHANK OR STALK)
Size and shape Is the stem thick, thin, short or noticeably tall?

Colour Is the stem the same colour as the cap, or is it different? Does it have coloured streaks? Does the colour differ at the base in relation to the colour nearer the apex, just before joining the cap?

Texture Is the stem smooth, scaly or of some other texture? Is it brittle, fibrous or flexible?

Ring (annulus) Is there a ring on the stem? Is it large, small, grooved or moveable? Does the stem have a different texture or colouring above and below it?

Markings Are there any distinct markings on the stem, such as pitting or freckling?

Consistency Is the stem brittle, fibrous or flexible?

Flesh Cut the stem. Is it hollow or solid? Is it darker near the base? Does the flesh change colour and if so, where? Is it clustered and/or joined with other mushrooms at the base?

Base Is the base thicker where it joins the substrate? Does it have a sack or volva (remnants of the universal veil)? Is the stem rooting?

2. HABITAT, SUBSTRATE AND TIME OF YEAR

Making notes about when and where you find any fungi is extremely relevant in the identification process.

HABITAT

The two main habitats are woodland and grassland, but also make a note of the specific types of these. Is the woodland broadleaved, coniferous or mixed? Is the grassland in an urban setting such as a garden or roadside? Is it in a meadow or field, or by a hedgerow?

SUBSTRATE

The base on which an organism lives, this is also a very important identification feature. Note the following:

Grassland Note where the fungus is growing, for example:
- On pasture land or a lawn, or in a meadow.
- In the soil or on dung, or some other medium.
- On woodchip, mulch, compost, sawdust or another medium.

Near woodland or trees Note where the fungus is growing, for example:
- Near a tree(s). If so, what species?
- In a clearing containing, for example, grass or leaf litter.
- On a living or dead tree. If so, what species?
- On dead wood, fallen branches, twigs or stumps.
- Near dead wood. Some species appear to grow from the soil but may in fact be feeding on dead wood or roots underground.

TIME OF YEAR

The main mushroom season begins in late summer and continues through to early winter, but many species grow earlier and/or persist late into the season. Many bracket fungi are perennial and winter is host to a few exclusive edible species.

3. HOW TO TAKE A SPORE PRINT

Knowing the colour of a spore deposit (reproductive spores dropped from the underside of the cap) is very useful for narrowing down identification. Simply cut off the mushroom stem as close as possible to where it joins the cap. Place the cap (gills down) on a sheet of blank white card. Leave undisturbed for 3–4 hours or overnight, then

Left: Black spore print.

remove the cap to see the results. Check the colour in daylight rather than under an electric light source. See the black spore print example above. You will have something very similar to this.

There are naturally many colours to be seen, such as white, cream, pink, brown and black, and shades of these in between. It is sometimes best to place the cap over black card, as paler spores will be more noticeable on a darker background.

If you do not wish to remove the stem, simply cut a hole in some cardboard, slot the stem through the hole and place the card on the rim of a tall glass, so that the mushroom is gently suspended on the card.

TO SUM UP

From all the information gathered, you will have a much better idea of what you are dealing with. Refer to these notes when cross-referencing with species featured in this guide and/or other sources.

WARNING/DISCLAIMER

There are many inedible and poisonous species of mushroom throughout Europe. A handful of these are deadly. Do not eat any fungus whose name you do not know, and if you are at all uncertain that you have identified a fungus correctly, do not eat it. The author accepts no liability for any injury or death occurring as a result of ingesting or exposure to any fungus described in this identification guide.

RUSSULAS

The Sickener
Russula emetica

The Sickener is a member of the very large *Russula* genus, commonly known as the brittlegills, the fruiting bodies and gills of which are brittle and fragile. This is one of the most common red-capped varieties, displaying a vibrant cherry red cap that does not fade with age, with a cuticle that can be easily peeled away from the margin to show reddening on the white flesh beneath. The delicate stem and brittle gills are also pure white. The fungus smells slightly fruity and has an unpleasant hot taste. It can cause severe sickness in some people if consumed.

KEY FEATURES Blood-red cap, domed then flattened.

 DIMENSIONS Cap 3–10cm; stem 4–9 x 1–2cm.

SPORE PRINT White.

HABITAT AND SEASON With pines; late summer–late autumn.

EDIBILITY Poisonous.

Left: The Sickener
Russula emetica.

Common Yellow Russula or Ochre Brittlegill
Russula ochroleuca

This species often grows in large numbers, even when conditions are dry. Its ochre-yellow cap is initially convex, soon flattening out and usually forming a central depression. The margin shows fine striations where it meets the creamy-coloured gills. There are two lookalike species that grow exclusively with beech and birch respectively. The Geranium Brittlegill *R. fellea* has a fruity odour, and the Yellow Swamp Brittlegill *R. claroflava* grows mostly in wetter conditions. Of these species only the Geranium Brittlegill, which has an unpleasant hot taste, is inedible.

KEY FEATURES Cap cuticle peels two-thirds up from margin.

 DIMENSIONS Cap 4–9cm; stem 4–8 x 1–2.5cm.

SPORE PRINT White or cream.

HABITAT AND SEASON With broadleaved trees and conifers; summer–autumn.

EDIBILITY Edible.

Common Yellow Russula *Russula ochroleuca*.

Crab Brittlegill
Russula xerampelina

This common European species can grow very large – the cap can reach up to 15cm across. The colour varies from dull red through to deep purple. When young the cap is small, stout and convex; it eventually flattens out and develops a central depression. One distinctive characteristic separates it from similar-coloured russulas. As the common English name suggests, it has a crab-like or fishy odour, which becomes more pungent with age. Despite this, it is a choice edible mushroom, albeit with a mild flavour. The offending smell soon disappears on cooking.

KEY FEATURES Fishy odour. Stem often tinted rose pink.

 DIMENSIONS Cap 5–15cm; stem 3–11 x 1–3cm.

SPORE PRINT Creamy-ochre.

HABITAT AND SEASON With broadleaved trees, especially beech and oak; late summer–late autumn.

EDIBILITY Edible. Mild.

Crab Brittlegill *Russula xerampelina*.

Cap has variable colouring.

Charcoal Burner
Russula cyanoxantha

The common name of this mushroom originates from the range of colours visible on its sturdy and fleshy cap, which are very much like the colours of a charcoal flame. It may sometimes just occur in one colour, but is often a variable mix of different hues such as blues, yellows, violets, greens and greys. It can be tricky to identify, but as an exception to the rule its gills are actually flexible, not rigid or brittle like those of other russulas. Running a finger across them will confirm this, so it is one distinguishing feature to look out for.

KEY FEATURES Variable colours. Flexible gills (not brittle).

 DIMENSIONS Cap 5–15cm; stem 5–10 x 1.5–3cm.

SPORE PRINT White.

HABITAT AND SEASON With broadleaved trees; summer–late autumn.

EDIBILITY Edible, mild.

Charcoal Burner *Russula cyanoxantha*.

MILKCAPS

Woolly Milkcap
Lactarius torminosus

Milkcaps exude a milk-like fluid from the gills and cap flesh when damaged; they occur in several colour variations, and these can sometimes change when exposed to air. The Woolly Milkcap is one of the few poisonous *Lactarius* mushrooms. Its salmon-coloured cap is initially convex and soon develops into a funnel shape with an inrolled margin that is covered in woolly hairs. The white-coloured milk that exudes from the gills has an unpleasant hot taste. Darker concentric bands on the cap are similar to those on the edible Saffron Milkcap (see p. 21), but this has orange milk.

KEY FEATURES Woolly hairs at margin.

 DIMENSIONS Cap 5–12cm; stem 4–8 x 1–2cm.

SPORE PRINT
Cream (yellowish).

HABITAT AND SEASON
Woods and parks, often with birch; late summer–early autumn.

EDIBILITY
Poisonous.

Woolly Milkcap
Lactarius torminosus.

18

Oakbug Milkcap
Lactarius quietus

Exclusive to oak woodland and parks, and small in size, the
Oakbug Milkcap is often overlooked due to its natural tawny-
brown colouring, which subtly blends into the environment. In
days gone by the smell was often described as being similar to
that of bed bugs. A modern, more relevant definition describes it
as bearing a similarity to light engine oil. The young cap is
rounded, and soon matures into a flatter shape with a distinctive
depressed centre. The cap surface is marked with darker
concentric bands and/or spots, which are often apparent but can
be subtle.

KEY FEATURES Small, in many numbers with oak. White milk.

 DIMENSIONS Cap 3–9cm; stem 4–9 x 1–1.5cm.

SPORE PRINT Cream, sometimes with a pinkish hue.

HABITAT AND SEASON With oak; autumn.

EDIBILITY Edible. Average.

Oakbug Milkcap *Lactarius quietus*.

Tawny Milkcap
Lactarius fulvissimus

The Tawny Milkcap has a long fruiting season. It is found in mixed woodland, appearing sporadically depending on location and moisture levels. It is predominately funnel-shaped with an orange or chestnut-brown colouring and distinctively paler margin. It exudes a white milk, which is plentiful albeit a little watery, and the flesh has an unpleasant smell. This milkcap is often found in small groups, favouring alkaline soils.

KEY FEATURES Orange or chestnut-brown to yellow-brown in colour.

 DIMENSIONS Cap 4–9cm; stem 3–7 x 0.5–2cm.

SPORE PRINT Cream.

HABITAT AND SEASON Mixed woodland; late summer–early winter.

EDIBILITY Inedible.

Tawny Milkcap *Lactarius fulvissimus*.

Saffron Milkcap
Lactarius deliciosus

This is a popular edible milkcap with a distinctive fleshy, orange-coloured cap. The surface features dark and light concentric bands. Darker pits or spots can be found on both the cap and the paler stem. It exudes orange milk that changes to a dull green within a day – consequently, older specimens often have green-stained caps and gills. The False Saffron Milkcap *L. deterrimus* and *L. semisanguifluus* have a similarly coloured milk, but are noticeably more reddish, turning deeper red or purplish over a 10–30-minute period. Neither species is poisonous.

KEY FEATURES Orange milk, fading to green.

 DIMENSIONS Cap 4–12cm; stem 3–6 x 1.5–2cm.

SPORE PRINT Pale ochre.

HABITAT AND SEASON With pines; summer–autumn.

EDIBILITY Edible. Very good.

Saffron Milkcap *Lactarius deliciosus*. Stem with spotted depressions.

WAXCAPS

Blackening Waxcap
Hygrocybe conica

As the scientific name suggests, the cap of this very common grassland mushroom is 'conical' in shape – usually very broad or bell-shaped. The texture, typical of all waxcaps, is smooth and greasy. The stem is yellow, often with a reddish hue, and the pale yellow, waxy gills have an adnexed to free attachment. Young specimens are bright with striking red, yellow and orange colours. Over time a 'blackening' process slowly takes effect, starting mainly from the cap edge and as streaks on the stem. Very old specimens turn completely black.

KEY FEATURES Bright colours. Turns black with age or handling.

 DIMENSIONS Cap 3–5.5cm; stem 3–7 x 0.5–1cm.

SPORE PRINT White.

HABITAT AND SEASON Grassland; autumn.

EDIBILITY Edible.

Blackening Waxcap *Hygrocybe conica*.   Blackened with age.

Parrot Waxcap
Hygrocybe psittacina

This is a small waxcap with variable colouring. The typical 'bell-shaped' cap develops a broad central umbo as it flattens out with age. It is initially green, but soon develops yellow- or sometimes pink-coloured patches. The stem shares the same colour variations, but is more dark blue-green towards the apex where it meets the cap and the yellow-green gills. The cap surface is covered in slimy green gluten, especially when the mushroom is young. In wet weather the whole fruiting body becomes extremely slimy and very hard to handle and take hold of.

KEY FEATURES Colourful greens and yellows. Slimy.

 DIMENSIONS Cap 1–3cm; stem 2–4 x 0.1–0.5cm.

SPORE PRINT White.

HABITAT AND SEASON Grass, heaths and gardens; summer–late autumn.

EDIBILITY Edible but too slimy.

Parrot Waxcap *Hygrocybe psittacina*.

Scarlet Waxcap or Scarlet Hood
Hygrocybe coccinea

The Scarlet Waxcap is a very attractive, deeply red-coloured mushroom. It is one of the larger waxcaps, growing up to 6cm across. There are several similar-looking species but this is by far the most common of the red variety, found exclusively in grassy fields. The cap is bell-shaped, in common with many *Hygrocybe* species, and has a noticeably slimy texture, especially when young. The flesh inside shares a similar colour but has a more yellowish hue. The stem often becomes slightly flattened or compressed, which is another good identification feature.

KEY FEATURES Striking red colour. Flattened stem.

 DIMENSIONS Cap 2–6cm; stem 2–5 x 0.3–1cm.

SPORE PRINT White.

HABITAT AND SEASON Grassy fields; late summer–autumn.

EDIBILITY Edible, pleasant.

Scarlet Waxcap *Hygrocybe coccinea*.

Butter Waxcap
Hygrocybe ceracea

There are several yellow/orange species of *Hygrocybe* and this is one of the smaller examples, only growing up to 2.5cm across. It has a warm yellow to deep orange colour throughout. The greasy, convex-shaped cap flattens out with age, and faint striate markings can be seen at the very edge. The thin and hollow stem often shows traces of fine white cottony down at the tapered base. Another characteristic to look out for in a close inspection lies under the cap, where the decurrent yellow gills are connected by tiny veins.

KEY FEATURES White down at base of stem. Gills joined by veins.

 DIMENSIONS Cap 1–2.5cm; stem 3–5 x 0.2–0.5cm.

SPORE PRINT White.

HABITAT AND SEASON Short grass in pastures and garden lawns; autumn.

EDIBILITY Edible.

Butter Waxcap *Hygrocybe ceracea*.

Underside, showing gills.

BONNETS AND MOSSCAPS

Angel's Bonnet
Mycena arcangeliana

Commonly known as bonnets, *Mycena* species are small and
delicate, bell- or conical-shaped mushrooms with long stems.
Angel's Bonnet can grow in large numbers on dead deciduous tree
stumps, branches and logs, of which it favours beech. It has a
whitish, broadly conical cap with distinctive grey-olive or pale
brown hues, and dark radiating striations. The thin stem is a
similar pale colour, often with whitish down at the base. Iodoform
is a naturally occurring compound in the flesh of the mushroom.
It has a strong, pungent smell, especially when crushed.

KEY FEATURES Striate markings. Chemical smell.

 DIMENSIONS Cap 1–5cm; stem 2–4 x 0.1–0.2cm.

SPORE PRINT Whitish.

HABITAT AND SEASON Stumps, logs and branches of deciduous
wood; autumn.

EDIBILITY Unknown. Possibly non-edible.

Angel's Bonnet *Mycena arcangeliana*.

Common Bonnet
Mycena galericulata

One of the larger bonnets, the Common Bonnet grows up to 6cm across when mature, with a broad central umbo. The species has mild brown colouring (darker near the centre) and a noticeable striated margin. It has white adnate gills, with a tiny decurrent tooth, eventually turning pink with age. The stem shares the same colour as the cap, but is clearly much lighter towards the apex where it meets the cap and gills. Common Bonnets are often found in large clusters on broadleaved tree stumps, branches and logs.

KEY FEATURES Broad cap, wide central bump (umbo).

 DIMENSIONS Cap 2–6cm; stem 2–10 x 0.3–0.9cm.

SPORE PRINT Cream.

HABITAT AND SEASON Stumps, logs and fallen branches of broadleaved trees; all year.

EDIBILITY Edible.

Common Bonnet
Mycena galericulata.

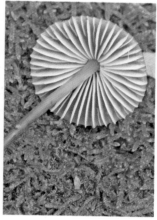

Gills eventually
turn pink.

Clustered Bonnet

Mycena inclinata

Identification of this mushroom is made relatively easy by the fact that it always appears in large, densely clustered groups, and only on oak stumps. The dark brown, bell-shaped caps have a prominent dark umbo, and the striated surface markings can easily be seen. At the margin there are tiny scalloped shapes from the cap that overhang the gills, which are another distinctive feature. The long stem is lighter at the apex and darker red-brown further towards the base, where the grouped stems are covered in a fine white down.

KEY FEATURES Margin. White down at base of stem.

 DIMENSIONS Cap 2–3cm; stem 5–10 x 0.2–0.4cm.

SPORE PRINT White.

HABITAT AND SEASON Oak stumps; late summer–autumn.

EDIBILITY Not edible. Unpleasant taste.

Clustered Bonnet *Mycena inclinata*.

Collared Mosscap
Rickenella swartzii (Mycena swartzii)

The Collared Mosscap is often found in damp or boggy grassland. It is frequently overlooked due to its tiny size. The small, creamy-brown cap has a very dark brown central marking. When young the cap is convex – it soon flattens out, eventually developing a central depression. The creamy-white decurrent gills meet the stem where there is a violet-coloured 'collar'. The rest of the stem is pale brown. The Orange Mosscap *R. fibula* is similar, but has a warmer, yellow-orange colour.

KEY FEATURES Violet colouring at stem apex. Dark centre in cap.

 DIMENSIONS Cap 0.5–1cm; stem 2–4 x 0.1–0.2cm.

SPORE PRINT White.

HABITAT AND SEASON Damp, short grass and moss; late summer–early winter.

EDIBILITY Not edible.

Collared Mosscap *Rickenella swartzii*.

FUNNELS

Clouded Agaric/Funnel
Clitocybe nebularis

Clitocybe species have funnel-shaped caps, and some species also have an umbo. The gills are very decurrent. The Clouded Agaric is large and fleshy, and is often found in large troops or rings. The cap is initially rounded but flattens out as it ages, eventually becoming deeply funnel-shaped. It is whitish with a 'cloudy' grey-brown colouring, especially at the centre, but is much lighter at the inrolled margin. The white gills are deeply decurrent and the stem is thick, fleshy and fibrous. The colouring becomes much lighter at the swollen base.

KEY FEATURES Cloudy brown-grey colours.

 DIMENSIONS Cap 5–20cm; stem 5–10 x 1.5–2.5cm.

SPORE PRINT Creamy-white.

HABITAT AND SEASON Coniferous and deciduous woodland; late summer–late autumn.

EDIBILITY Edible. Good, but can cause stomach upsets.

Clouded Agaric/Funnel
Clitocybe nebularis.

Stem swollen at base.

Common Funnel
Clitocybe gibba

The Common Funnel is mainly found in scattered groups on deciduous woodland floors, in grassy clearings and on heaths. Its smooth, pale ochre-coloured cap is funnel-shaped throughout its life and often has a wavy margin, especially when older. The crowded white decurrent gills attach themselves at an acute angle to the strong and distinctly cylindrical stem, which is often swollen at the base. Although the species is edible, it is best left alone as it can be confused with several similar-looking poisonous *Clitocybe* mushrooms.

KEY FEATURES Sweet bitter almond smell.

 DIMENSIONS Cap 3–8cm; stem 3–8 x 0.5–1cm.

SPORE PRINT White.

HABITAT AND SEASON Deciduous woodland and grassy patches; summer–late autumn.

EDIBILITY Edible, but can be confused with poisonous species so best left alone.

Common Funnel *Clitocybe gibba*.

Aniseed Funnel
Clitocybe odora

A unique and attractive *Clitocybe*, the Aniseed Funnel is often found in the leaf litter of mixed deciduous woodland, favouring beech. Is also occurs in coniferous woodland on clay soils. The obvious difference between this and other funnels lies in the blue-green colouring and strong aniseed odour of the flesh. The smell is more of a useful identification marker than the colour, which can fade or become dulled. The stem is paler than the cap and is always covered in a fine white down.

KEY FEATURES Blue-green colours. Aniseed smell.

 DIMENSIONS Cap 3–7cm; stem 3–6 x 0.5–1cm.

SPORE PRINT White.

HABITAT AND SEASON Deciduous and coniferous woodland; summer–late autumn.

EDIBILITY Edible. Dried and used as seasoning.

Aniseed Funnel *Clitocybe odora*.

Fool's Funnel
Clitocybe rivulosa

This species is one of the smaller funnels that can be found, and like most of the smaller species it is poisonous and should be avoided – several deaths following ingestion of the mushroom have been recorded. It is commonly found in rings or small groups in parks, and on grassland, paths and roadsides. The young, small white cap soon expands and develops a slight central depression. It is covered in a fine white powder, giving it a frosty appearance, and has pale pinkish markings where the flesh underneath shows through. The decurrent gills are compact and range in colour from white to buff. The mushroom has a faint sweetish odour.

KEY FEATURES Small and white with fleshy patches.

 DIMENSIONS Cap 2–5cm; stem 2–4 x 0.4–1cm.

SPORE PRINT White.

HABITAT AND SEASON Grass, paths, roadsides and disturbed ground; summer–autumn.

EDIBILITY Deadly poisonous.

Fool's Funnel *Clitocybe rivulosa*.

Powdered white cap.

KNIGHTS

Grey Knight
Tricholoma terreum

Knight mushrooms are medium to large agarics with varying cap textures. Most are dull brown or grey, with a few paler orange-yellow exceptions. The Grey Knight is medium sized, and as its common name suggests has a light to dark grey cap with distinctive soft, downy fibres on its surface, giving it a felt-like texture. The stem is white (sometimes discoloured) and very smooth, becoming hollow with age. The gills are slightly spaced apart and are notched just before they attach to the stem (known as an 'emarginate attachment').

KEY FEATURES Felty-textured cap. Emarginate gills.

 DIMENSIONS Cap 4–7cm; stem 3–8 x 1–1.5cm.

SPORE PRINT White.

HABITAT AND SEASON All woodland, especially Scots Pine; late summer–late autumn.

EDIBILITY Edible.

Grey Knight *Tricholoma terreum*.

Girdled Knight
Tricholoma cingulatum

This is one of the few *Tricholoma* species with a ring on the stem, albeit the ring is small and almost non-distinct at times. Its grey cap flattens out at maturity and has a subtle central bump; it is covered in fine grey-brown fibres that have a felt-like texture. The white stem is quite long with a slight woolly ring and the gills, which are also white, have an emarginate attachment. The flesh has a strong smell of flour (meal) and this is also passed on to the taste, which is not to everyone's liking.

KEY FEATURES Small, woolly ring. Grey, felted cap.

 DIMENSIONS Cap 3–6cm; stem 5–8 x 0.8–1.2cm.

SPORE PRINT White.

HABITAT AND SEASON Meadows, with willows; late summer– late autumn.

EDIBILITY Edible.

Girdled Knight *Tricholoma cingulatum*.

TRICHOLOMA ALLIES

Plums and Custard
Tricholomopsis rutilans

This agaric is a very aptly named mushroom – it is named after its natural colouring. It can grow very large, a typical average cap size being around 8cm. The purple colour effect is created by many purple-reddish flecks or scales on a predominately yellow cap; these are usually densest at the centre. The same colour features on the stem, but the scales here are less profuse. The distinctive rich yellow gills do have an uncanny 'custard' hue. The species is always found growing on and around conifer stumps and logs. Some people consider it edible, while others find it too earthy and bitter.

KEY FEATURES Purple-reddish and yellow colouring.

 DIMENSIONS Cap 7–12cm; stem 4–6 x 1–2cm.

SPORE PRINT White.

HABITAT AND SEASON Near and on conifer stumps; late summer–late autumn.

EDIBILITY Edible. Not recommended.

Plums and Custard
Tricholomopsis rutilans.

Distinctive purple and yellow colouring.

Common Cavalier
Melanoleuca polioleuca

Cavalier is the common name for mushrooms in the *Melanoleuca* genus. They all feature wide, flattish caps with a central raised bump or umbo. The Common Cavalier is one of the most common and has a smooth-textured, distinctly deep brown cap that fades to pale buff as it dries. The flattened cap has a slight depression in the centre where the typical umbo can clearly be seen. Other important identification features include the streaky brown fibres covering the creamy-white stem, which is quite bulbous at the base. The gills are sinuate and creamy-white.

KEY FEATURES Central bump in a depressed cap.

 DIMENSIONS Cap 3–8cm; stem 4–7 x 0.8–1.5cm.

SPORE PRINT White/cream.

HABITAT AND SEASON Woods and pastures; late summer–late autumn.

EDIBILITY Edible. Not recommended.

Common Cavalier
Melanoleuca polioleuca.

Typical central bump (umbo).

Porcelain Fungus
Oudemansiella mucida

The common name of this species originates in its ivory-white, semi-translucent cap, although it is more greyish when young. The surface of the cap is covered in a slimy film that is most apparent in moist conditions. The species is parasitic, growing in large groups, always on beech trees and also on dead branches, logs and stumps. The white stem has striate markings above the small, flimsy ring, and is slightly scaly below. When growing out of the side of a tree the stems bend to position the caps upright. The gills are pure white and distant with an adnate attachment.

KEY FEATURES Slimy. Semi-translucent cap.

 DIMENSIONS Cap 2–8cm; stem 3–10 x 0.3–1cm.

SPORE PRINT White.

HABITAT AND SEASON Beech trees and dead wood; late summer–early winter.

EDIBILITY Edible. Remove glutinous film.

Porcelain Fungus *Oudemansiella mucida*.

Cap is semi-translucent and slimy.

Rooting Shank
Xerula radicata

This noticeably tall mushroom grows near deciduous trees, especially under beech. The stem roots downwards under the soil and attaches to the roots (or dead wood) of the tree. The taller the mushroom, the longer the root, which can make it very difficult to remove from the ground in one piece. The cap is initially rounded and flattens out at maturity; it features a persistent umbo with radiating wrinkles, creating a smooth, bumpy surface. The cap colour ranges from pale ochre to dark olive-brown. The surface texture is always slimy, but especially so after wet weather, and the gills are white and moderately spaced.

KEY FEATURES Rooting stem.

DIMENSIONS Cap 3–10cm; stem 8–20 x 0.5–1cm.

SPORE PRINT White.

HABITAT AND SEASON Near deciduous trees; early summer–late autumn.

EDIBILITY Edible but uninteresting.

Rooting Shank
Xerula radicata.

Cap is slimy with
radiating wrinkles.

CALOCYBE

St George's Mushroom
Calocybe gambosa

This is one of the few good edible mushrooms that can be found early in the season, around St George's Day (23 April) to be precise, but it usually fruits a few weeks later. It occurs in all types of grassland, including pastures and woodland edges. The cap, stem and crowded gills are all white in colour, and the flesh has a floury or cucumber-like odour (mealy). The Deadly Fibrecap (see p. 87) appears at around the same time and may be confused with this species, but it is confined to deciduous woodland and path edges, and it reddens with age.

KEY FEATURES Spring season. Flour or cucumber odour.

 DIMENSIONS Cap 6–15cm; stem 2–4 x 1–2.5cm.

SPORE PRINT White.

HABITAT AND SEASON Grassland, pastures and wood edges; April–May.

EDIBILITY Edible. Very good.

St George's Mushroom *Calocybe gambosa*.

Honey Fungus or Bootlace Fungus *Armillaria mellea*.

HONEY FUNGI

Honey Fungus or Bootlace Fungus
Armillaria mellea

The Honey Fungus is one of the most destructive of all parasitic fungi, causing white rot and spreading from tree to tree via long, black stringy cords known as rhizomorphs, which resemble bootlaces. It grows in dense and plentiful clusters on and around trunks and stumps of all tree types in woodland, parks and gardens. The concave, ochre-brown cap flattens out with age and is covered in fine dark scales. The long, tapering stem has a high-positioned ring and features some yellow colouring at the edge.

KEY FEATURES Dark cap fibres. Yellow edge on ring.

 **DIMENSIONS** Cap 4–12cm; stem 6–15 x 0.5–1.5cm.

SPORE PRINT Pale cream.

HABITAT AND SEASON On and around deciduous and coniferous trees and stumps in woodland, parkland, gardens and orchards; summer–early winter.

EDIBILITY Edible. Must be cooked well.

Old 'Bootlace' cords; rhizomorphs.

Dark Honey Fungus
Armillaria ostoyae

The Dark Honey Fungus is very similar looking to the Honey Fungus (p. 43), but with several subtle differences. It grows in similar clustered groups, during the same season and in the same habitats, but it does favour spruce woodland. The cap is a darker ochre-brown and features the same dark, fibrous scales that are denser at the centre. The yellow-white stem becomes red-brown towards the slightly swollen base. The ring has a distinctive dark brown edge, positioned further down the stem when compared to the common Honey Fungus.

KEY FEATURES Dark edge to ring.

 DIMENSIONS Cap 4–15cm; stem 6–15 x 0.5–1.5cm.

SPORE PRINT White/cream.

HABITAT AND SEASON On and around deciduous and coniferous trees and stumps; summer–early winter.

EDIBILITY Edible. Must be cooked well.

Dark Honey Fungus *Armillaria ostoyae*.

DECEIVERS

The Deceiver
Laccaria laccata

The *Laccaria* (or the deceivers) are a genus of mushroom whose cap colour and shape can vary dramatically. They often grow in large trooping groups in an assortment of shades and sizes. The cap is frequently wavy with striated markings at the margin. Its colours range from reddish-brown when moist to pale yellow-buff when dry, and its texture is usually fine and scurfy. The stem is tough and fibrous, and often twisted, and its soft pink-brown gills are widely spaced, often dusted white with spores.

KEY FEATURES Tough stem. Widely spaced gills.

 **DIMENSIONS** Cap 1.5–6cm; stem 4–10 x 0.5–1cm.

SPORE PRINT White.

HABITAT AND SEASON Mixed woodland and heaths; summer– early winter.

EDIBILITY Edible. Little taste.

The Deceiver *Laccaria laccata*.

Distinctive widely spaced gills.

Amethyst Deceiver
Laccaria amethystina

The Amethyst Deceiver is almost identical in every way to the common Deceiver (see p. 45), except for the purple/lilac colouring. Once a few of the mushrooms have been discovered, many more will be found in the immediate vicinity. Typical *Laccaria* features such as the mature flattened wavy cap, striate margin and widely spaced gills are all present. Depending on weather conditions and age, the colour shades often vary – the species is normally deep purple in moist conditions and very pale lilac when dry. The cap features a fine scurfy texture and the stem is covered in fine white fibres.

KEY FEATURES Purple colouring. Widely spaced gills. Tough stem.

 DIMENSIONS Cap 1.5–6cm; stem 4–10 x 0.4–0.8cm.

SPORE PRINT White.

HABITAT AND SEASON Mixed woodland, especially beech; late summer–early winter.

EDIBILITY Edible. Little taste.

Amethyst Deceiver
Laccaria amethystina.

Gills dusted white from spores when mature.

PARACHUTES AND MARASMIUS

Fairy Ring Champignon
Marasmius oreades

This is a popular edible mushroom that is sold throughout European markets, although it is unpopular with many gardeners. The mushrooms are often found in full or partial rings where the grass slowly fades and dies from the fringe inwards, yet flourishes at the point where the mushrooms grow, in the nutrient-rich soil. The cap is yellow-brown when wet and dries to a pale straw-like colour. It features a distinctive and persistent central umbo, and a subtly striated margin. The stem remains pale whitish and the gills, which are initially white, age to an ochre-cream and are widely spaced.

KEY FEATURES Large, broad umbo. Grows in rings.

 DIMENSIONS Cap 2–5cm; stem 2–10 x 0.3–0.5cm.

SPORE PRINT White.

HABITAT AND SEASON Grass in pastures, lawns and verges; spring–autumn.

EDIBILITY Only cook the caps.

Right: Fairy Ring Champignon *Marasmius oreades*.

Partial 'ring' in the grass.

Horsehair Parachute
Marasmius androsaceus

Fungi of the *Marasmius* genus are commonly known as the parachutes due to their caps often resembling parachute canopies. Most of them are very small, as in this case, so they are often overlooked. The Horsehair Parachute is found on dead needles, leaves and twigs, and dead heather. Its cap is pinky-brown, often darker at the centre, and is noticeably wrinkled with a central depression. The gills are similar in colour and widely spaced, and the black stem is tall and extremely thin (almost hair-like).

KEY FEATURES Very small. Long, thin black stem.

 DIMENSIONS Cap 0.5–1cm; stem 2–6cm x 0.1cm.

SPORE PRINT White.

HABITAT AND SEASON On twigs, leaves and needles; late spring–late autumn.

EDIBILITY Not edible. Unpleasant.

Horsehair Parachute *Marasmius androsaceus*.

Collared Parachute
Marasmius rotula

The Collared Parachute grows on dead twigs, roots and fallen leaves. A key feature of this species is that the gills are free from the stem and centrally connect to a fleshy collar. The tiny white 'parachute-like' cap is noticeably ridged or scalloped at the margin, and often has a dark, depressed centre. The stem is very dark but whiter at the apex, and the gills are very widely spaced. Even though the mushrooms are very small and delicate, they often occur in large trooping groups scattered on debris across the woodland floor.

KEY FEATURES Gills attached to collar on stem.

 DIMENSIONS Cap 0.5–1.5cm; stem 2–7 x 0.1cm.

SPORE PRINT White.

HABITAT AND SEASON Dead roots, twigs and leaves; summer–winter.

EDIBILITY Not edible. Unpleasant.

Collared Parachute *Marasmius rotula*.

TOUGHSHANKS AND SIMILAR SPECIES

Toughshank or Spindleshank
Collybia fusipes

Toughshank is the common name for this genus of mushroom due to their tough, fibrous and flexible stems. This species fruits early in the year, and is found exclusively at the bases of deciduous trees and stumps, often in large, clustered groups. The red-brown caps are rounded when young, expanding slightly at maturity. The species has the added characteristic of a twisting, central, flattened stem. The base is tapered and roots into the soil, and is whitish in colour, darkening red-brown towards the base. The gills are free with a small decurrent 'toothed' attachment to the stem.

KEY FEATURES In groups. Twisted, compressed stem.

 DIMENSIONS Cap 3–7cm; stem 4–9 x 0.7–1.5cm.

SPORE PRINT White.

HABITAT AND SEASON Bases of deciduous trees and stumps; spring–early winter.

EDIBILITY Inedible. Too tough.

Toughshank or Spindleshank
Collybia fusipes.

Always in dense groups.

Spotted Toughshank
Collybia maculata

This mushroom favours coniferous woodland and can often be hidden from view under bracken in woods or on heathland. Initially the clean creamy-white caps are dome-shaped; they then flatten out, developing tawny-brown freckles as they age. These spots sometimes merge together to create larger brown patches, especially near the centre. The white, crowded and free gills, along with the tough, flexible stem, also become speckled tawny-brown with age. The long stems often have rooting bases.

KEY FEATURES Spotted brown markings, especially with age.

 DIMENSIONS Cap 4–10cm; stem 5–10 x 0.8–1.2cm.

SPORE PRINT Cream–pale pink.

HABITAT AND SEASON Deciduous and coniferous woods, and heaths under bracken; summer–late autumn.

EDIBILITY Edible. Bitter and tough.

Spotted Toughshank *Collybia maculata*.

Brown freckled cap.

Russet Toughshank
Collybia dryophila

This is a variable, small brown mushroom that grows in leaf and needle litter in all types of woodland, usually in small groups scattered about the woodland floor. The cap colour ranges from very pale tan to pale brown-yellow or orange-brown. The margin is often wavy, more so with age. The gills are free to annexed, and whitish to pale yellow-brown. The species' common name derives from its tough reddish-brown stem, which is darker towards the slightly bulbous base. Tearing the stem apart reveals the hollow interior.

KEY FEATURES Tough reddish-brown stem, bulbous at base.

 DIMENSIONS Cap 2–5cm; stem 2–4 x 0.2–0.4cm.

SPORE PRINT White or pale creamy-pink.

HABITAT AND SEASON Leaf litter in all woodland types; late spring–late autumn.

EDIBILITY Edible. Not the best.

Russet Toughshank
Collybia dryophila.

Butter Cap
Collybia butyracea

This extremely common toughshank grows in abundance at the height of the mushroom season. It can be found scattered on the ground in both deciduous and coniferous woodland. The cap has a shallow dome shape that flattens out with age, and features a prominent umbo. It has a distinctive butter-like texture that is very smooth and slippery. In colour it ranges from dark red-brown and buff to ivory. The gills are crowded, free from the stem and remain whitish. The stem shares the same colour as the cap and becomes thicker at the base, which is covered in fine white down. It is tough, stringy and hollow.

KEY FEATURES Slippery, smooth cap.

 DIMENSIONS Cap 3–8cm; stem 3–6 x 0.5–1.2cm.

SPORE PRINT White.

HABITAT AND SEASON Deciduous and coniferous woods; autumn–early winter.

EDIBILITY Edible.

Butter Cap *Collybia butyracea.*

Velvet Shank
Flammulina velutipes (Collybia velutipes)

The Velvet Shank is of the few mushroom species that appear during the winter months. It is a very popular edible species, especially because it grows when not many other mushrooms can be found. It can survive harsh frosts, reappearing once conditions are milder and appearing in grouped clusters on dead deciduous trees and stumps. The cap is orange-brown, darker at the centre, and has a smooth, slimy texture, and the gills are pale yellow. The key characteristic of this species lies in the stem, which is extremely tough, is dark brown and lighter at the apex, and has a distinctly velvet-like texture.

KEY FEATURES Tough, dark velvety stem.

 DIMENSIONS Cap 2–10cm; stem 3–10 x 0.4–0.8cm.

SPORE PRINT White-cream.

HABITAT AND SEASON Decaying deciduous trees and stumps; late autumn–spring.

EDIBILITY Edible and good.

Velvet Shank *Flammulina velutipes*.

FALSE CHANTERELLE

False Chanterelle
Hygrophoropsis aurantiaca

This mushroom is aptly named because of the common confusion
between it and the true Chanterelle (see p. 129). All parts of the
fruiting body are orange with misleading decurrent gills. The
species is said to be edible but a few people have suffered from
hallucinations after ingesting it and it is therefore best avoided.
The main differences between the two species are that the cap of
the False Chanterelle has a fine downy texture and tends to be
much more orange in places than that of the true Chanterelle. It
has gills rather than false, thick, fleshy blades, and lacks the
apricot odour of the true Chanterelle. The spore colour of the
False Chanterelle is also white rather than ochre-yellow.

KEY FEATURES   Yellow-orange. Fine, felty cap.

 DIMENSIONS   Cap 2–8cm; stem 3–5 x 0.5–1cm.

SPORE PRINT   White.

HABITAT AND SEASON   Coniferous woods and heaths; autumn.

EDIBILITY   Inedible.

False
Chanterelle
*Hygrophoropsis
aurantiaca*.

PARASOLS

Parasol Mushroom
Macrolepiota procera

The relatively large species in the *Macrolepiota* genus are commonly known as parasols. The Parasol Mushroom, named because of its resemblance to a parasol umbrella, initially has an egg-shaped, white-buff cap covered with large dark brown scales. It soon expands, sometimes to as large as 25cm across. The scales spread and regularly disperse, leaving a prominent and dark central bump. The long, thin stem has a snakeskin pattern and features a large double ring that can be moved up and down the stem. The gills are white and free.

KEY FEATURES Snakeskin stem. Moveable ring.

 DIMENSIONS Cap 10–25cm; stem 15–35 x 1.5–2.5cm.

Parasol Mushroom *Macrolepiota procera*.

Cap can be up to 25cm across.

SPORE PRINT White.

HABITAT AND SEASON Grassland, parks, verges and open woodland; summer–autumn.

EDIBILITY Edible, nutty flavour.

Shaggy Parasol
Macrolepiota rhacodes

Often mistaken for the Parasol Mushroom (see p. 56), the Shaggy Parasol is usually thickset with a stockier appearance, and is more often found with conifers. The rounded white cap expands to almost flat with age, and has dark brown scales that curve upwards to create the shaggy appearance. The base of the stem is thick and rounded, unlike that of the Parasol Mushroom, which is not as bulbous; however, the ring can similarly be moved up and down the stem. The flesh, including the white gills, bruises red-orange when handled or cut.

KEY FEATURES Shaggy scales. Red bruising.

 DIMENSIONS Cap 12–18cm; stem 8–12 x 1.5–2.5cm.

SPORE PRINT White.

HABITAT AND SEASON Woodland, parks and gardens; summer–autumn.

EDIBILITY Edible. Some people have mild reactions, such as gastric upset or rash, after eating these mushrooms.

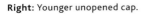

Shaggy Parasol
Macrolepiota rhacodes.

Right: Younger unopened cap.

DAPPERLINGS

Stinking Dapperling
Lepiota cristata

Most dapperlings are unknown for their edibility and several are poisonous or even deadly, so they are best avoided for consumption. The Stinking Dapperling is extremely common, and is found in all types of woods, on lawns and in garden waste. It often grows in small, scattered groups on the ground, and has a wide, bell-shaped cap with a distinctive scaly red-brown umbo. From here the scales break up and thinly disperse across the white cap towards the margin. The flesh has a pungent, unpleasant odour, hence the species' common name.

KEY FEATURES Unpleasant smell. Red-brown central umbo.

 DIMENSIONS Cap 2–5cm; stem 2–4 x 0.2–0.4cm.

SPORE PRINT White.

HABITAT AND SEASON Woods and gardens; summer–autumn.

EDIBILITY Possibly poisonous. Avoid.

Stinking Dapperling *Lepiota cristata*.

Freckled Dapperling
Lepiota aspera

Similar in many ways to the Stinking Dapperling (see p. 59), the Freckled Dapperling has dark brown scales covering a larger creamy-white cap, which are most prominent at the centrally raised bump. Smaller conical scales spread across the rest of the cap, producing a freckled effect. The cap is elliptical when young, and soon develops into a bell shape; it eventually becomes much flatter while still retaining a prominent umbo. There is a small ring on the white, pallid-brown stem that is slightly cotton-like in consistency, and the free white gills are forked at the point where they meet the stem.

KEY FEATURES Dark brown central umbo.

 DIMENSIONS Cap 5–10cm; stem 3–5 x 0.5–1cm.

SPORE PRINT White.

HABITAT AND SEASON Deciduous woods; autumn.

EDIBILITY Unknown. Avoid.

Freckled Dapperling *Lepiota aspera*.

AMANITAS

Fly Agaric
Amanita muscaria

The Fly Agaric is one of the most recognisable of all mushrooms and is depicted in many illustrations featured in books about fairytales. Its striking red cap is covered in the white scaly remnants of a universal veil that initially encapsulates the young mushroom, appearing egg-like in shape. As the mushroom grows the veil breaks apart, leaving behind these distinctive markings. The pure white stem has a large basal bulb with warted surface markings. The gills are also pure white, crowded and free. Avoid eating this poisonous mushroom as it causes sickness and has the potential to be fatal.

KEY FEATURES Red colour, white scales. Bulbous, warted stem base.

 DIMENSIONS
Cap 10–20cm; stem 15–20 x 1.5–2cm.

SPORE PRINT White.

HABITAT AND SEASON With birch, pine and spruce; late summer–early winter.

EDIBILITY Hallucinogenic. Can be fatal.

Fly Agaric *Amanita muscaria*.

Younger, unopened cap.

Panthercap *Amanita pantherina.*

Panthercap
Amanita pantherina

With the typical appearance of many *Amanita* species, the
Panthercap features a large, bulbous stem base and spotted cap.
In this case the cap colour is ochre-brown with pure white warty
veil remnants. The stem and free gills are also pure white. A key
identification feature lies on the stem. Unlike similar-looking
species such as the Grey Spotted Amanita and Blusher (see p. 64
and 65), which are edible if cooked, the Panthercap's ring has no
grooves on the upper side. It is poisonous and should not be
consumed.

KEY FEATURES Ring has no grooves.

 DIMENSIONS Cap 5–10cm; stem 6–10 x 1–2cm.

SPORE PRINT White.

HABITAT AND SEASON Broadleaved and coniferous woods;
summer–autumn.

EDIBILITY Poisonous. Can be fatal.

The cap is covered in pure white veil remnants.

Grey Spotted Amanita or False Panthercap
Amanita excelsa (var. spissa)

The Grey Spotted Amanita has two key defining characteristics setting it apart from its poisonous cousin, the Panthercap (see p. 63). The ochre-brown cap is covered in off-white or greyish veil remnants, rather than pure white ones, and there are noticeable grooved markings on the upper side of the ring. Other features, such as the bulbous stem base and white gills, are not reliable for distinguishing between the two species. This mushroom is said to be edible after cooking, but always take great care in identification.

KEY FEATURES Grey scales. Ring is striated.

DIMENSIONS Cap 10–15cm; stem 10–12 x 1–2cm.

SPORE PRINT White.

HABITAT AND SEASON Broadleaved and coniferous woods; summer–autumn.

EDIBILITY Edible if cooked well. Avoid if unsure of identification.

Grey Spotted Amanita or False Panthercap *Amanita excelsa*.

Left: Ring on stem has striated markings.

The Blusher
Amanita rubescens

Once cooked, this *Amanita* is edible and good, but it is similar in appearance to the poisonous Panthercap (see p. 63). The Blusher has the usual characteristics of most amanitas, featuring a large, fleshy brown cap with scaly veil remnants, a thick basal bulb, free white gills and a large membranous ring. The ring has distinctive grooved markings on its upper side, but a decisive identification marker is that the flesh turns rosy-red once it has been bruised, damaged or cut. The cap surface colour, which ranges from light ochre-brown to rosy-brown, can often have a reddish hue, especially in older specimens.

KEY FEATURES Red bruising. Grooved markings on upper ring.

DIMENSIONS Cap 5–15cm; stem 6–14 x 1–2.5cm.

SPORE PRINT White.

HABITAT AND SEASON Coniferous and deciduous woods; summer–autumn.

EDIBILITY Edible. Cook well. Avoid if unsure of identification.

The Blusher *Amanita rubescens*.

Deathcap *Amanita phalloides*.

Deathcap
Amanita phalloides

The Deathcap is one of the most poisonous mushroom species in the world, containing deadly amatoxins that cause extreme sickness in the victim, who will temporarily recover, only to then go into liver and kidney failure, and eventually die. Treatment, but no cure, is available if the symptoms of poisoning are caught early. The Deathcap's medium to large flattened cap, which is initially rounded, has faint radiating streaky markings. The colour is often yellowish-green or olive tinted. It can also just be a flush of these colours. The white, often banded stem features a large basal bulb that is encased in a sack-like volva. The gills are crowded, free and pure white.

KEY FEATURES Olive/yellow-green colouring. Large volva.

 DIMENSIONS Cap 5–12cm; stem 10–12 x 1–2cm.

SPORE PRINT White.

HABITAT AND SEASON Mixed deciduous woodland; summer–autumn.

EDIBILITY Deadly poisonous.

The cap features very faint radiating streaks.

Destroying Angel
Amanita virosa

This is an attractive pure white mushroom with deadly potential. The symptoms caused after consumption are very similar to those caused by ingesting the Deathcap (see p. 67). *Amanita* species most often have a distinctive basal bulb, sometimes in a bag-like enclosure called a volva that is the remnant of a protective veil. In this case the stem has a moderately small, swollen base surrounded by an off-white casing. The ring is often torn and fragmented, and the cap, which is initially rounded or bell-shaped, expands into a shallow dome with a sticky consistency, especially after rain. All parts of the mushroom are pure white.

KEY FEATURES Pure white. Bulbous stem in volva.

 DIMENSIONS Cap 4–9cm; stem 5–7 x 1–1.5cm.

SPORE PRINT White.

HABITAT AND SEASON Coniferous and deciduous woodland; late summer–autumn.

EDIBILITY Deadly poisonous.

Destroying Angel
Amanita virosa.

Grisette
Amanita vaginata

The Grisette is a particularly slender *Amanita* species that lacks a ring on the stem or any veil remnants on the cap, although it is encased in a large white-grey volva, typical of so many other *Amanita* mushrooms. The grey-brown cap is ovate or half-egg-shaped at first, then expands to a flatter shape featuring a slightly darker and shallow umbo. The cap is noticeably striated at the margin. The tall whitish stem, which is tinted with the cap colour, thins out towards the cap where the white, crowded gills have an adnexed attachment.

KEY FEATURES Bag-like volva. Light striated margin.

 DIMENSIONS Cap 5–9cm; stem 13–20 x 1.5–2cm.

SPORE PRINT White.

HABITAT AND SEASON Deciduous woods and heaths; summer–autumn.

EDIBILITY Edible. Cook well.

Grisette *Amanita vaginata*.

Young, closed cap.

BLEWITS

Wood Blewit
Lepista nuda

The Wood Blewit is one of the few edible mushrooms to be found during winter, although it does appear from mid-autumn onwards. Occurring in all types of woodland, it also appears in hedgerows and gardens. The cap is usually thick, fleshy and tough, with a blue-lilac hue. These colours fade to brownish or pale buff with age. The crowded, sinuate gills remain violet for longer, but fade eventually. The pale and fibrous stem is often marked with strong lilac streaks, so is a more reliable identification feature.

KEY FEATURES Lilac gills. Lilac-streaked stem.

 DIMENSIONS Cap 5–15cm; stem 5–10 x 1.5–2.5cm.

SPORE PRINT Pale pink.

HABITAT AND SEASON Woods and gardens; late autumn–winter.

EDIBILITY Edible, good. Cook well.

Wood Blewit *Lepista nuda*.

Above: *L. nuda* can survive winter snow and frosts.

Left: Lilac-coloured gills eventually fade to buff.

Field Blewit

Lepista saeva

Similar in some ways to the Wood Blewit (see p. 70), the Field Blewit, as its name suggests, is found in fields and woodland edges, often in rings. The mature specimen has a thick, fleshy cap that becomes wavy with an inrolled margin. The colour varies from pale yellow-brown to grey-brown, and the thick, fibrous stem has a bulbous base and is covered in strong violet- or lilac-coloured streaks. The crowded, sinuate gills are quite pale, often whitish to light ochre or flesh-coloured.

KEY FEATURES Violet streaks on stem.

 DIMENSIONS Cap 5–12cm; stem 4–10 x 1.5–2.5cm.

SPORE PRINT Pale pink.

HABITAT AND SEASON Grassland and woodland edges; autumn–winter.

EDIBILITY Edible and good. Cook or dry.

Field Blewit *Lepista saeva*.

Tawny Funnel
Lepista flaccida

The Tawny Funnel is often confused with funnel mushrooms (*Clitocybe*), but for several scientific reasons it has been moved to the *Lepista* genus, known as the blewits. It is a medium-sized mushroom maturing into a distinctive funnel shape with deeply decurrent whitish-yellow gills. The cap surface is orange or yellow-brown, and there are often water markings at the edge of the cap. The tawny colouring darkens as the mushroom ages. The stem is paler than the cap and becomes hollow after a while. The base is covered in fine white woolly fibres.

KEY FEATURES Often grouped. Orange-brown. Funnel shape.

 DIMENSIONS Cap 5–9cm; stem 2.5–5 x 0.5–1cm.

SPORE PRINT White.

HABITAT AND SEASON Deciduous and coniferous woods, in leaf litter; summer–early winter.

EDIBILITY Edible, but not good.

Tawny Funnel *Lepista flaccida*.

Right: Often found in groups.

PINKGILLS AND RELATED SPECIES

Shield Pinkgill
Entoloma clypeatum

Entoloma species are commonly known as the pinkgills. Many have interesting smells and their gills turn to shades of pink after the spores mature. The Shield Pinkgill grows exclusively with trees and shrubs of the *Rosaceae* (rose) family, often in small groups or rings. The greyish-brown to dark brown cap is initially convex, then flattened with a broad umbo, and features faint radiating lines. The contrasting white stem is tinted with the cap colour and has fine streaky white markings. The gills are initially pale grey, and mature to pale pink.

KEY FEATURES Grows with *Rosaceae*.

 DIMENSIONS Cap 3.5–10cm; stem 3.5–5.5 x 0.7–1.5cm.

SPORE PRINT Pink.

HABITAT AND SEASON With rose, hawthorn and cherry and other rose family members; spring–midsummer.

EDIBILITY Not edible. Avoid.

Shield Pinkgill *Entoloma clypeatum*.

Grey gills eventually turn pink.

Star Pinkgill
Entoloma conferendum

The common name of this species refers to its star-shaped spores when seen under a microscope. It is a very common mushroom that is found in both grassland and open woodland. Its small, convex cap is hygrophanous, and changes colour due to water content in the cap tissue. In fair conditions it is brown, reddish or grey-brown. It becomes pallid as it dries and has a noticeably striate margin when moist. The flesh has a floury smell and the thin, pale stem is covered in fine, whitish fibres. The white gills later mature to pink.

KEY FEATURES Fine white stem fibres. Floury smell.

 DIMENSIONS Cap 1–3cm; stem 2–6 x 0.1–0.3cm.

SPORE PRINT Pink.

HABITAT AND SEASON Grassland and open woodland; autumn.

EDIBILITY Not edible.

Star Pinkgill *Entoloma conferendum*.

The Miller
Clitopilus prunulus

The Miller is a choice edible mushroom, but care must be taken not to confuse it with poisonous *Clitocybe* species. All parts of the mushroom are white in colour except the deeply decurrent gills, which turn from white to pink as the spores mature. The smooth cap has a texture reminiscent of chamois leather. It is rounded when young, but soon expands, becoming irregularly wavy with an inrolled margin. The short stem often has an off-centre attachment to the cap and the flesh has a strong smell of fresh flour (meal).

KEY FEATURES Strong floury smell.

DIMENSIONS Cap 3.5–10cm; stem 2–4 x 0.4–1.5cm.

SPORE PRINT Pink.

HABITAT AND SEASON Grass in open woodland or parks; summer–late autumn.

EDIBILITY Edible and good.

The Miller *Clitopilus prunulus*.

SHIELDS

Deer Shield
Pluteus cervinus

The *Pluteus* genus, commonly known as shield mushrooms, has a pink spore print just like that of the entolomas (see p. 74 and 75), and grows exclusively on dead wood and wood debris. The Deer Shield is one of the most common species, growing on dead deciduous tree stumps and trunks. The dark brown cap is rounded when young and broadly convex at maturity. Darker brown radiating streaks can be seen on the surface. The contrasting white stem, which is swollen at the base, is covered in long dark brown fibres, and the free white gills turn pink as the spores mature.

KEY FEATURES Dark brown cap with radiating streaks.

 DIMENSIONS Cap 4–12cm; stem 7–11 x 0.5–1.5cm.

SPORE PRINT Pink.

HABITAT AND SEASON Rotting deciduous tree trunks and stumps; summer–late autumn.

EDIBILITY Edible.

Deer Shield *Pluteus cervinus*.

Velvet Shield or Veined Pluteus
Pluteus umbrosus

The Velvet Shield is an attractive species of *Pluteus*, growing on rotting deciduous wood. It features a distinctive sepia-brown, velvet-textured cap with darker radiating fibres on raised wavy ridges, giving it a 'veined' appearance. The central area appears darker due to a denser concentration of these fibres, which also appear on the white stem, covering the entire surface. The gills are free and initially white, eventually changing to pink as the spores mature (typical of many *Pluteus* species). On closer inspection of the gills a distinct brown edge can be seen.

KEY FEATURES Velvet-like texture. Radiating brown scales.

 DIMENSIONS Cap 3–9cm; stem 3–10 x 0.5–1.3cm.

SPORE PRINT Pink.

HABITAT AND SEASON Rotting deciduous wood; autumn.

EDIBILITY Edible.

Velvet Shield or Veined Pluteus *Pluteus umbrosus*.

WEBCAPS

Deadly Webcap
Cortinarius rubellus

Webcaps are a large genus, split into many subgenera. Most are difficult to identify and are either inedible or poisonous. The common name for the group derives from the cobweb-like attachment from the cap to the stem, especially when the mushrooms are young. The Deadly Webcap has been confused with the edible Chanterelle (see p. 129), so it is essential to know the superficial differences. The cap, which is initially convex, expands to a shallow bell shape; it is tawny to red-brown and covered in fine scaly fibres. The paler stem is also fibrous and covered in patchy cap colours. The pale ochre adnate gills mature to rust-brown.

KEY FEATURES Spore print colour.

 DIMENSIONS Cap 2.5–8cm; stem 5–11 x 0.5–1.5cm.

SPORE PRINT Rust-brown.

HABITAT AND SEASON Coniferous woods, in moss; autumn.

EDIBILITY Deadly poisonous.

Deadly Webcap *Cortinarius rubellus*.

PAXILLUS

Brown Rollrim
Paxillus involutus

This extremely common and poisonous toadstool is still sold as an edible mushroom in eastern Europe. Some toxicity is lost after cooking, but if consumed regularly over many years the poison can build up and attack the red blood cells, which can be fatal. The cap has a hazel-brown colour, tawny-olive brown when young, often dotted with darker orange-brown blotches. When young, the texture is finely felted; it later becomes smooth and slimy when wet. The rim remains inrolled and the crowded, decurrent gills bruise dark brown on handling and easily separate from the cap.

KEY FEATURES Inrolled rim. Brown bruising to gills.

 DIMENSIONS Cap 5–15cm; stem 7–8 x 0.8–1.3cm.

SPORE PRINT Yellowish-brown.

HABITAT AND SEASON With deciduous and coniferous trees, in parks and gardens; late summer–late autumn.

EDIBILITY Poisonous. Can be deadly.

Brown Rollrim *Paxillus involutus*.

RUSTGILLS

Common Rustgill
Gymnopilus penetrans

The Common Rustgill's name originates from the rust-coloured spores deposited from the gills. Sometimes the spores can be seen on the caps of mushrooms, where mushrooms above them have dropped spores onto them. The Common Rustgill is always found in conifer woods and usually grows in large groups. Its smooth, rich yellow-brown cap is flat and often wavy at the margin. The yellow-brown stem is covered in fine white fibres and has a woolly base. The slightly decurrent gills are also yellow-brown and become spotted with brown as they age.

KEY FEATURES White fibres on brown stem. Woolly base.

DIMENSIONS Cap 3–8cm; stem 4–7 x 0.5–1cm.

SPORE PRINT Rust.

HABITAT AND SEASON On dead conifer wood and debris; summer–late autumn.

EDIBILITY Bitter. Avoid.

Common Rustgill
Gymnopilus penetrans.

SCALYCAPS

Golden Scalycap
Pholiota aurivella

The scalycaps are often found in dense groups on stumps, trunks and other woodland debris. The Golden Scalycap is so named because its bell-shaped/convex cap has a strong golden-yellow colouring. There are small dark brown scales scattered across the surface, making it appear spotted. The texture is slightly sticky or slimy when moist. The stem often has an off-centre attachment; it is usually bent with a scraggy texture and is often missing a ring. The gills are adnate and crowded, and pale yellow maturing to orange-brown.

KEY FEATURES Orange-brown cap with dark-spotted scales.

 DIMENSIONS Cap 5–15cm; stem 5–8 x 0.5–1.5cm.

SPORE PRINT Rust-brown.

HABITAT AND SEASON Hardwood, conifer logs and trunks; summer–autumn.

EDIBILITY Inedible but non-poisonous.

Golden Scalycap *Pholiota aurivella*.

Shaggy Scalycap
Pholiota squarrosa

The Shaggy Scalycap is usually found in large, visually striking groups. The medium to large cap is quite pale or straw-yellow, and is covered in thick brown, upturned scales. The long, often curved stem also has many scales, which become smaller and finer towards the darkening base. Apart from on the crowded cinnamon-brown gills, which are pale yellow when young, the only smooth area to be found is just above the torn ring zone, very close to where it meets the cap.

KEY FEATURES Scaly stem and cap.

 DIMENSIONS Cap 4–11cm; stem 5–12 x 1–1.5cm.

SPORE PRINT Rust-brown.

HABITAT AND SEASON Bases of deciduous and coniferous trees; late summer–autumn.

EDIBILITY Inedible. Bitter.

Shaggy Scalycap *Pholiota squarrosa*.

Shaggy Scalycap found at the base of deciduous and coniferous trees.

POISONPIES

Poisonpie
Hebeloma crustuliniforme

One of the most common *Hebeloma* species, the Poisonpie is a small to medium-sized mushroom found in open woodland, and on park paths and verges under deciduous trees. Its broadly convex to flattened cap is an off-white, pale ochre colour, and is often greasy or viscid when wet. The free, clay-brown gills often 'weep' with small droplets of water when conditions are wet; this feature is unique to only a few species. The species also gives off a strong smell of radish, which becomes more apparent if the flesh is crushed. The stem is mostly white or off-white in colour.

KEY FEATURES Strong radish smell.

 DIMENSIONS Cap 4–10cm; stem 4–7 x 1–2cm.

SPORE PRINT Rust.

HABITAT AND SEASON Under deciduous trees, in open woodland; autumn.

EDIBILITY Poisonous. Avoid.

Poisonpie
Hebeloma crustuliniforme.

Sweet Poisonpie

Hebeloma pallidoluctosum

Very similar in many respects to the common Poisonpie (see p. 85), the Sweet Poisonpie is also found in woodland, but in more 'damp' situations, scattered across the woodland floor. It is similar in size to its common cousin, and the greasy off-white to brown cap is convex to flattened, often with a shallow umbo. It is more ochre-brown in the centre and the majority of the whitish stem has a silky, fibrous texture, less so nearer the apex. The gills are initially pale brown, and soon mature to a rust-brown colour.

KEY FEATURES Smells sweet and flowery.

 DIMENSIONS Cap 2–7cm; stem 4–8 x 0.5–1.3cm.

SPORE PRINT Dark rust.

HABITAT AND SEASON Damp woodland; autumn.

EDIBILITY Inedible. Avoid.

Sweet Poisonpie
Hebeloma pallidoluctosum.

Flowery, sweet odour
from the flesh.

FIBRECAPS AND SIMILAR SPECIES

Deadly Fibrecap or Red-staining Fibrecap
Inocybe erubescens

Inocybe species are mostly dull and overlooked, and many are difficult to identify. Their caps are usually streaky and fibrous. Many are inedible or poisonous, and at least one, like the species featured here, is deadly. The Deadly Fibrecap is found in deciduous woods, usually beech, often at the sides of paths. When young it has been known to be confused with St George's Mushroom (see p. 41), but the mature, ivory-coloured, bell-shaped cap has fibres that discolour red as the mushroom ages. The rosy-red gills turn to cream and eventually olive-brown. They also turn red with age or on bruising.

KEY FEATURES Fibrous. Stains red.

 DIMENSIONS Cap 2–8cm; stem 3–10 x 0.5–1.2cm.

SPORE PRINT Brown.

HABITAT AND SEASON Deciduous woods; spring–autumn.

EDIBILITY Deadly poisonous.

Deadly Fibrecap *Inocybe erubescens*.

Fool's Conecap
Conocybe filaris

Fungi in the *Conocybe* genus can be similar in appearance to *Inocybe* species, and are often difficult to identify. They generally have cone-shaped caps in a range of dull brown colours. Fool's Conecap is a slight exception to the rule, having a more rounded or bell-shaped cap. Its colour ranges from creamy-white to pale tan. The white stem soon fades to a yellowish colour and features a small white membranous ring. With age, the ring becomes stained orange-red from the deposited spores above. The adnate gills are tan or yellow-brown.

KEY FEATURES Small striate ring. White turning rust.

 DIMENSIONS Cap 0.5–1.5cm; stem 3–5 x 0.1–0.3cm.

SPORE PRINT Yellow-brown rust.

HABITAT AND SEASON Grass; autumn.

EDIBILITY Inedible.

Fool's Conecap
Conocybe filaris.

GALERINA AND TUBARIA

Dwarf Bell
Galerina pumila

Many *Galerina* species are quite small, as in this case, and are commonly found growing in moss, often in gardens and damp grassland. As its common name suggests, the Dwarf Bell has a conical to broadly bell-shaped cap. It has strong yellow-brown colouring and is noticeably striate at the margin, especially in wet weather. Over time the colour fades to pale yellow. The delicate stem is the same colour as the cap, and is covered in fine white fibres when young. The gills have an adnate attachment and are paler than the cap.

KEY FEATURES Striated yellow-brown cap.

DIMENSIONS Cap 0.5–2cm; stem 3–6 x 0.1–0.2cm.

SPORE PRINT
Ochre.

HABITAT AND SEASON Moss; late summer–late autumn.

EDIBILITY Not edible.

Dwarf Bell
Galerina pumila.

Scurfy Twiglet
Tubaria furfuracea

The Scurfy Twiglet is a common, small brown mushroom that can be found all year round, most often in autumn or early spring. Its cap is ochre-brown; it is initially conical but soon expands to a flatter shape. There are faint striated markings at the margin, which are more noticeable when the cap is moist. When dry it has a mildly flaky or scurfy texture, often with remnants of white velum. The stem is similarly coloured and is covered in a white cotton-like down at the base. The slightly decurrent gills are yellow-brown in colour.

KEY FEATURES Scurfy dry cap. White velum markings.

 DIMENSIONS Cap 1–4cm; stem 2–5 x 0.2–0.5cm.

SPORE PRINT Pale yellow-brown.

HABITAT AND SEASON Wood debris; summer–autumn.

EDIBILITY Not edible.

Scurfy Twiglet *Tubaria furfuracea*.

WOODTUFTS AND BROWNIES

Sheathed Woodtuft or Velvet Toughshank
Kuehneromyces mutabilis

This popular edible mushroom grows in large clustered groups on deciduous stumps and trunks. Care must be taken not to confuse it with the deadly Funeral Bell *Galerina marginata*. The convex to flattened cap is changeable in appearance. It is bright orange-brown when moist through to pale ochre as it dries from the centre of the cap outwards, producing a two-toned effect. The stem is a key identification marker. It is dark brown and scaly below its small dark ring, and smooth and pale above. The gills are initially pale, and cinnamon-brown when mature.

KEY FEATURES Scaly below ring, smooth above.

DIMENSIONS Cap 3–6cm; stem 3–8 x 0.5–1cm.

SPORE PRINT Dark ochre.

HABITAT AND SEASON Trunks and stumps of deciduous trees; spring–late autumn.

EDIBILITY Edible. Good.

Sheathed Woodtuft *Kuehneromyces mutabilis*.

Sulphur Tuft
Hypholoma fasciculare

Hypholoma species are also commonly known as Brownies and the larger species are commonly known as Tufts. Sulphur Tuft is a common, year round toadstool, usually in large fruiting displays on trunks or stumps. The mature rounded cap is bright sulphur-yellow with darker orange tones at the centre. Remnants of the pale yellow veil often cover the margin, and the fibrous stem is usually curved. It is similar in colour to the cap, although it may sometimes be greenish with a dark ring zone. It becomes darker brown towards the base, and is often fused with other stems. The young gills are greenish-yellow and are a good identification marker; they eventually mature to dark brown.

KEY FEATURES Large groups. Yellow-green gills.

 DIMENSIONS Cap 2–7cm; stem 4.5–10 x 0.5–1cm.

SPORE PRINT Dark purple-brown.

HABITAT AND SEASON Dead deciduous or coniferous wood; all year.

EDIBILITY Poisonous.

Sulphur Tuft *Hypholoma fasciculare*.

Sulphur Tuft grows in tight, crowded clusters.

Brick Tuft
Hypholoma lateritium

The Brick Tuft is similar in many ways to the Sulphur Tuft (see p. 92), growing in large groups on tree stumps or logs. However, it is only ever found on deciduous wood and the stems are usually longer than those of the Sulphur Tuft. The cap is brick-red at the centre; it becomes paler towards the margin, where there are often reddish-brown veil remnants still attached. The stems are pale yellow at the apex and orange-brown at the base, where they often merge together. The pale yellow gills turn to olive-brown as the spores mature.

KEY FEATURES Brick-red cap with veil remnants.

 DIMENSIONS Cap 3–11cm; stem 5–17 x 0.5–1.3cm.

SPORE PRINT Dark purple-brown.

HABITAT AND SEASON Deciduous tree stumps and thick wood; autumn.

EDIBILITY Edible but bitter.

Brick Tuft *Hypholoma lateritium*.

MUSHROOMS

Field Mushroom
Agaricus campestris

Agaricus simply translates as mushroom. The genus features the well-known Supermarket Mushroom (see p. 99). Wild-growing agarics such as the Field Mushroom are often superior in flavour. This species grows naturally in fields and meadows at least 20m away from any trees, which helps separate it from any other similar poisonous *Agaricus* mushrooms. It often grows in rings. The young white, domed cap soon expands to a broadly convex shape, leaving little or no ring behind on its white stem. The texture can be smooth or slightly scaly, and the dark pink gills of the young specimen mature to dark brown.

KEY FEATURES Location.

DIMENSIONS Cap 3–10cm; stem 3–10 x 1–2cm.

SPORE PRINT Brown.

HABITAT AND SEASON Meadows and fields; late summer–autumn.

EDIBILITY Edible. Very good.

Field Mushroom
Agaricus campestris.

Above: Field Mushroom cap can be smooth or slightly scaly.

Right: Young pink gills.

The Prince
Agaricus augustus

Favouring coniferous woodland, The Prince is typically large with distinctive brown-flecked scales on its cap. When young the whole fruiting body is rounded, stocky and thick. The veiled white gills display dotted markings close to the margin before the ring drops and hangs from the white stem. Small scales are present below the ring, and they turn browner with age. As it grows, the cap becomes broadly convex, its scales spread apart and the gills turn brown. Its flesh has a strong almond smell and develops patchy red hues as it ages.

KEY FEATURES Almond, marzipan odour. Brown scales.

 DIMENSIONS Cap 10–25cm; stem 10–20 x 2–4cm.

SPORE PRINT Deep purple-brown.

HABITAT AND SEASON Coniferous and deciduous woods; summer–autumn.

EDIBILITY Edible. Very good.

The Prince
Agaricus augustus.

Pavement Mushroom or Spring Agaricus
Agaricus bitorquis

Roadsides and verges are commonplace locations for this urban mushroom – it is quite a remarkable species, with its power to push through very dense soil and even tarmac. The species is pure white, and is sometimes covered in very pale, ochre-coloured scales. It has a short, thick stem and rounded cap that expands flatter, and is often covered in soil or asphalt remnants. The double ring on the stem (one curving upwards and the other downwards) is a distinctive feature to look out for. The gills are dark pink when young, then clay-brown and finally dark brown.

KEY FEATURES Location. Double ring.

 DIMENSIONS Cap 4–10cm; stem 3–6 x 1.5–2cm.

SPORE PRINT Brown.

HABITAT AND SEASON Roadsides, verges and gardens; late spring–autumn.

EDIBILITY Edible. Very good.

Pavement Mushroom *Agaricus bitorquis*.

Cultivated Mushroom
Agaricus bisporus

This is the mushroom most of the Western world is familiar with. Many variations at different stages of maturity are sold in most supermarkets. The wild variety can look very different, due to its darker cap – many fine brown fibres radiate around the pale ochre cap. The cap is initially rounded but soon expands as it grows. The cut or damaged flesh has a natural 'mushroomy' smell and bruises pinkish-red when cut or damaged. The stem is often flaky below the floppy membranous ring, and the young dirty pink gills mature to dark brown.

KEY FEATURES Brown scaly cap. Flesh bruises reddish.

 DIMENSIONS Cap 5–10cm; stem 8–10 x 2–4cm.

SPORE PRINT Chocolate-brown.

HABITAT AND SEASON Dung heaps, compost and rich soil; late spring–autumn.

EDIBILITY Edible. Very good.

Cultivated Mushroom *Agaricus bisporus*.

Horse Mushroom
Agaricus arvensis

This is a tasty edible mushroom that grows in most types of grassland and often in rings. The common name refers to its large size. The pure white cap is initially rounded, and can expand to up to 20cm across; it is always smooth. Before the ring appears on the thick white stem, the gills are covered by a white veil that features a distinctive 'cogwheel' pattern on the outer circumference. The young gills are white at first, then turn pink and eventually chocolate-brown. The flesh has a pleasant mushroom smell with a hint of aniseed. It develops pale yellow, patchy colouring as it ages.

KEY FEATURES Cogwheel pattern; veiled gills.

 DIMENSIONS Cap 7–20cm; stem 8–10 x 2–3cm.

SPORE PRINT Deep purple-brown.

HABITAT AND SEASON Grassland and pasture; autumn.

EDIBILITY Edible. Very good.

Horse Mushroom *Agaricus arvensis*.

Cogwheel pattern on the Horse Mushroom protective veil.

Mature specimens of Horse Mushroom can grow up to 20cm across.

Yellow Stainer
Agaricus xanthodermus

Responsible for many cases of food poisoning, the Yellow Stainer is often confused with Horse and Field Mushrooms (see p. 100 and 95). While some people suffer from gastric upset after eating it, others are not affected. The similarities between the species are superficial, but there are some noticeable differences. Both the Horse Mushroom and the Yellow Stainer 'bruise' yellow, but there is a distinctly more chromium-yellow colour to the latter, especially on the rim of the cap when rubbed, and on the flesh inside the base of the stem. Other differences include an unpleasant phenol or inky smell, a large, floppy ring and white gills when very young.

KEY FEATURES Chrome-yellow staining.

 DIMENSIONS Cap 5–15cm; stem 5–15 x 1–2cm.

SPORE PRINT Deep purple-brown.

HABITAT AND SEASON Woods, gardens, meadows and hedgerows; summer–autumn.

EDIBILITY Poisonous. Can cause sweating, stomach cramps and sickness.

Yellow Stainer
Agaricus xanthodermus.

Cap margin stains
'chrome yellow.'

Wood Mushroom

Agaricus silvicola

Care must be taken not to confuse this edible mushroom with some of the deadly poisonous *Amanita* species (see p. 61–69). Although the Wood Mushroom often has a swollen base at the end of the stem, it lacks a volval sack, and at no stage in its development does it have white gills. The young white cap is bell-shaped, and soon expands to broadly convex with thin flesh. It usually develops yellow colouring with age or when damaged. The ring is three-quarters of the way up the stem, and is large and floppy. The flesh has a pleasant aniseed odour.

KEY FEATURES Large ring. Bulbous stem.

 DIMENSIONS Cap 5–10cm; stem 5–9 x 1–1.5cm.

SPORE PRINT Deep purple-brown.

HABITAT AND SEASON Coniferous and deciduous woods; autumn.

EDIBILITY Edible. Good.

Wood Mushroom *Agaricus silvicola*.

FIELDCAPS

Yellow Fieldcap
Bolbitius titubans

The Yellow Fieldcap can be found in various locations, but often occurs in well-manured grass. The young cap is bright chrome-yellow, and initially rounded or ovate; it expands to a bell shape and eventually becomes flat with deep grooves at the margin. The flesh is fragile and thin, and almost translucent and glutinous when wet. With age, it pales to yellow-ochre. The delicate, hollow stem is white-yellow, and is covered in a very fine white powder. The crowded, pale yellow gills characteristically turn brown to rust-brown with age.

KEY FEATURES Chrome-yellow striated cap. Spore colour.

 DIMENSIONS Cap 1–4cm; stem 4–10 x 0.2–0.4cm.

SPORE PRINT Rust-brown.

HABITAT AND SEASON Manured grass, rotting straw, dung and woodchips; summer–late autumn.

EDIBILITY Not edible.

Yellow Fieldcap *Bolbitius titubans*.

Bright yellow cap when very young.

Common Fieldcap
Agrocybe pediades

The Common Fieldcap can be a tricky mushroom to identify as it falls into the awkward category of 'small and brown' with no real outstanding features. It grows in varied locations such as grass on roadsides and in gardens, but can also be found on sandy dunes. The cap is small, yellow-brown and almost flat. The stem is very thin and paler than the cap, with darker cap-coloured markings here and there. The adnate gills are moderately spaced apart and pale yellow-brown; they eventually turn dark brown with age as the spores mature.

KEY FEATURES Mealy smell.

 DIMENSIONS Cap 1–3cm; stem 2.5–4 x 0.2–0.5cm.

SPORE PRINT Dark brown.

HABITAT AND SEASON Grass in gardens and verges, and sand dunes; summer–autumn.

EDIBILITY Edible. Uninteresting.

Common Fieldcap *Agrocybe pediades*.

ROUNDHEADS

Verdigris Agaric
Stropharia aeruginosa

Roundhead is the common name for the *Stropharia* genus, of which the Verdigris Agaric is a particularly stunning and conspicuous member. It is often found in small groups in woods and grassland. The domed, deep green-blue cap is sporadically covered in persistent white-flecked scales, and has a sticky, glutinous texture. Over time, pale yellow-white patches appear randomly on the surface. The stem is similarly coloured and more densely coated in the fine white scales below the ring. The whitish gills eventually turn purple-brown with age and often have a white edge.

KEY FEATURES Colouring. White colour at gill edge.

 DIMENSIONS Cap 2–8cm; stem 4–10 x 0.4–1.3cm.

SPORE PRINT Brown-purple.

HABITAT AND SEASON Woods, grassland and soil; early summer–late autumn.

EDIBILITY Not edible.

Verdigris Agaric *Stropharia aeruginosa*.

Blue Roundhead
Stropharia caerulea

Similar to the *Verdigris Agaric* (see opposite), the Blue Roundhead is also blue-green in colour, albeit slightly paler, and is much more common. Cap and stem features are very similar too, so identification can be difficult. Tiny white fibres are initially present at the cap margin but soon disappear. The pallid stem has a slightly fibrous and scaly texture below the small, short-lived ring, and is smoother above. The gills are pale vinaceous-brown, turning darker brown with age, and lack the white edge feature of *S. aeruginosa*.

KEY FEATURES Blue-green cap discolours to pale yellow-white.

 DIMENSIONS Cap 3–8cm; stem 4–10 x 0.4–1.3cm.

SPORE PRINT Dark yellow-brown.

HABITAT AND SEASON Leaf litter and grass near trees; summer–autumn.

EDIBILITY Not edible. Unpleasant.

Blue Roundhead *Stropharia caerulea*.

PSILOCYBE

Magic Mushroom or Liberty Cap
Psilocybe semilanceata

The Magic Mushroom is famous for its hallucinogenic properties, and it is now illegal to possess it in most countries. It contains the active compound psilocybin, which produces the psychedelic effects. The species can often be found growing in abundance in grassy meadows, fields and gardens. The yellowish-brown cap is elongated and conical, and often has a small, protruding 'nipple'. It is sticky and striated when moist, and fades to a much paler colour when dry. The whitish-yellow stem is relatively long and often bruises blue on handling. The pale brown gills mature to a much darker purple-brown.

KEY FEATURES Nipple often present on cap.

 DIMENSIONS Cap 0.5–1.5cm; stem 3–8 x 0.1–0.3cm.

SPORE PRINT Purple-brown.

HABITAT AND SEASON Grass in lawns and fields, and on paths; summer–autumn.

EDIBILITY Hallucinogenic. Similar-looking *Psilocybe* species can be deadly.

Magic Mushroom *Psilocybe semilanceata*.

BRITTLESTEMS

Common Stump Brittlestem
Psathyrella piluliformis

As with all *Psathyrella* species, the stems (and caps) are very
fragile and easily breakable. The Common Stump Brittlestem
often fruits early in the year in dense clusters at the bases of
deciduous trees. The young, smooth caps are orange-brown and
distinctly rounded, often with cottony white veil remnants at the
margin. Over time they expand to broadly convex, usually
discolouring paler with a darker centre, creating a two-toned
effect. This all depends on weather conditions and moisture
levels. The pale brown gills are crowded and eventually mature to
deep chocolate-brown.

KEY FEATURES In crowded groups. Orange-date colouring.

DIMENSIONS Cap 1.5–3cm; stem 4–9 x 0.5–1cm.

SPORE PRINT
Dark brown.

**HABITAT AND
SEASON**
Woodland on
stumps and thick
wood; late spring–
late autumn.

EDIBILITY
Edible. Bitter.

Common Stump
Brittlestem
*Psathyrella
piluliformis*.

Pale Brittlestem
Psathyrella candolleana

This pale ochre mushroom soon dries to whitish-grey, sometimes with a brownish hue. It is initially conical, and expands to almost flat, sometimes with a slight umbo. It often has white fibrous veil remnants scattered over the cap, especially overhanging the margin, and pale grey-violet colouring on the crowded adnate gills. As the spores mature the gills turn a deep chocolate-brown. The mushrooms often grow in small, clustered groups, and the brittle white stems are hollow, merging together at the base.

KEY FEATURES Cobweb-like veil remnants on cap.

 DIMENSIONS Cap 2–6cm; stem 4–9 x 0.3–0.8cm.

SPORE PRINT Dark brown.

HABITAT AND SEASON Near deciduous trees and stumps, and in grassy verges; late spring–late autumn.

EDIBILITY Edible. Mild when young.

Pale Brittlestem *Psathyrella candolleana*.

Clustered Brittlestem
Psathyrella multipedata

Many *Psathyrella* species grow in clumped dense groups, but none more so than the Clustered Brittlestem. It is often found in grass on path edges and roadsides, and in open woodland. The pale grey-brown caps are closely packed together and feature a paler, faintly striated margin. With age, they fade to a pallid grey-cream colour. The creamy-white stems meet and fuse together at a common base. The spore print produced from the crowded adnate and white-edged gills is very dark brown with a deep purple hue.

KEY FEATURES Large crowded groups.

DIMENSIONS Cap 1–3cm; stem 7–12 x 0.2–0.5cm.

SPORE PRINT Dark purple-brown.

HABITAT AND SEASON Grass in open woodland, roadsides and verges; late spring–summer.

EDIBILITY Not edible.

Clustered Brittlestem *Psathyrella multipedata*.

WEEPING WIDOW

Weeping Widow
Lacrymaria lacrymabunda

The Weeping Widow is edible, albeit bland and slightly bitter. The cap is rounded or broadly convex with a distinct umbo; it is variable in colour, ranging from pale brown to dark ochre-brown. Fine, woolly fibres cover its surface, which becomes smoother and darker with age. The white stem has a cotton-like ring, which is usually discoloured dark brown from fallen spores. Water droplets collect on the dark purple-brown gills, a characteristic that is referred to as 'weeping', hence the species' common name.

KEY FEATURES Fine, woolly texture. 'Weeping' gills.

 DIMENSIONS Cap 2–10cm; stem 4–8 x 0.5–1.2cm.

SPORE PRINT Dark purple-brown.

HABITAT AND SEASON In grass and soil; late spring–late autumn.

EDIBILITY Edible. Bitter.

Weeping Widow
Lacrymaria lacrymabunda.

Water droplets often found between gills.

INKCAPS

Shaggy Inkcap or Lawyer's Wig
Coprinus comatus

All inkcaps have dark-coloured spores and some often produce a black inky fluid as they mature. The shaggy appearance of this mushroom is due to the white, brown-tipped scales covering its tall, elliptically shaped cap. Young specimens have a pure white cap, often with light brown colouring at the apex. With age, the cap opens slightly and the gills (initially white, then pink) slowly blacken and auto-digest (deliquesce). Often there is little left but an inky mass at the top of a very long stem.

KEY FEATURES Shaggy appearance. Inky fluid.

 DIMENSIONS Cap 5–15cm; stem 8–15 x 1–2cm.

SPORE PRINT Dark brown-black.

HABITAT AND SEASON In grass on lawns or in disturbed areas; late summer–autumn.

EDIBILITY Edible. Young specimens are good.

Shaggy Inkcap or Lawyer's Wig
Coprinus comatus.

Gills turn to ink with age.

Common Inkcap *Coprinopsis atramentaria*.

Common Inkcap
Coprinopsis atramentaria

Similar in many ways to the Shaggy Inkcap (see p. 113), the Common Inkcap features the same elliptical or egg-shaped cap, but with greyish-brown colours and a smooth texture, often peppered with dark black-brown scales at the centre. Over time the greyish white gills turn black, eventually deliquescing. The mushrooms appear in small clustered groups on grass or soil, growing off dead wood below the surface. Even though the species is edible, it causes severe sickness if consumed with alcohol. Abstaining from drinking alcohol for 48–72 hours before and after consumption is recommended.

KEY FEATURES Elliptical smooth cap. In groups.

 DIMENSIONS Cap 4–8cm; stem 6–17 x 0.8–1.5cm.

SPORE PRINT Black.

HABITAT AND SEASON From buried wood under soil and grass; spring–late autumn.

EDIBILITY Edible. Bad reactions if mixed with alcohol.

Often growing in small dense groups.

Magpie Inkcap
Coprinopsis picacea

Aptly named due to its contrasting black-and-white markings, the Magpie Inkcap is one of the tallest growing fungi in the genus, sometimes reaching up to 25cm in height. The young, elliptically shaped cap is pale brown-grey, darkening to black as it expands and becoming bell-shaped; it is covered in greyish-white veil remnants all over the surface. Eventually, as is the case with many inkcaps, the gills dissolve to an inky fluid. The stem is pure white and covered in very fine white fibres, and has a slightly bulbous, woolly base.

KEY FEATURES Grey-black cap with white veil remnants.

 DIMENSIONS Cap 5–8cm; stem 9–25 x 0.5–1.5cm.

SPORE PRINT Black.

HABITAT AND SEASON Deciduous woods, often with beech; late summer–autumn.

EDIBILITY Not edible. May be poisonous to some.

Magpie Inkcap
Coprinopsis picacea.

White veil remnants on black cap.

Hare's Foot Inkcap
Coprinopsis lagopus

As it first emerges from the substrate, this mushroom displays an uncanny resemblance to a hare's foot. The elongated grey-brown cap is covered in whitish veil remnants that have the appearance of fine fur. The species often grows in scattered groups on disturbed soil, bark mulch and woodchip piles. As the cap expands, the white downy stem grows very tall, losing its fleecy fibres, and becomes broadly bell-shaped with distinctive radial striations. The gills soon deliquesce, and the cap becomes flatter and translucent, and often curls upwards at the margin.

KEY FEATURES Fur-like covering when young.

DIMENSIONS Cap 2–4cm, and 4–6cm across when mature; stem 6–13 x 0.2–0.4cm.

SPORE PRINT Black.

HABITAT AND SEASON Soil, mulch and woodchips; summer–autumn.

EDIBILITY Edible.

Hare's Foot Inkcap
Coprinopsis lagopus.

Mature, tall specimens.

Fairy Inkcap or Trooping Crumble Cap
Coprinellus disseminatus

The Fairy Inkcap is usually found in exceptionally large groups, sometimes in the hundreds, on and around deciduous tree stumps and their hidden roots. The small, conical to bell-shaped caps are yellow-beige with a darker tawny-coloured centre (maturing to grey-beige), and are packed closely together. From almost the top of the cap downwards they have noticeably deep grooves or striations. The smooth stem is whitish-beige, and slightly more yellowish towards the downy base. The white gills mature to black but do not turn into an inky fluid.

KEY FEATURES Strongly striated caps. Very large groups.

 DIMENSIONS Cap 0.5–1.5cm; stem 1.4–3.5 x 0.1–0.3cm.

SPORE PRINT Dark ochre-brown.

HABITAT AND SEASON
Deciduous tree stumps and hidden roots; late spring–late autumn.

EDIBILITY
Edible. Insubstantial.

Fairy Inkcap
Coprinellus disseminatus.

Glistening Inkcap
Coprinellus micaceus

This small, ovate inkcap usually grows in large numbers on or around broadleaved tree stumps and trunks and buried wood. It is similar to the Fairy Inkcap (see opposite), but is noticeably larger with the distinctive feature of being covered in fine, mica-like scales or flecks. The cap colour ranges from mild ochre to warm orange-brown; it is paler yellow at the striated margin and dark brown at the apex. The smooth stem is white, as are the young gills, but these soon mature to dark brown, then black, before eventually deliquescing.

KEY FEATURES Fine, mica-like flecks on the cap.

 **DIMENSIONS** Cap 1–4cm; stem 1.5–4 x 0.2–0.4cm.

SPORE PRINT Dark brown-black.

HABITAT AND SEASON On and around broadleaved tree stumps; late spring–late autumn.

EDIBILITY Edible.

Glistening Inkcap
Coprinellus micaceus.

Mica-like flecks on cap.

Firerug Inkcap
Coprinellus domesticus

The Firerug Inkcap is superficially like the Glistening Inkcap (see p. 119), having a similar ovately shaped cap, but it is paler ochre in colour with a tawny-brown centre, and the sprinkled white veil remnants are noticeably larger. The cap becomes smooth with age, and typical striated markings appear from the margin upwards. The smooth white stem is slightly bulbous at the base and often grows from a reddish-brown, fibrous mat of mycelium, a feature unique to this particular inkcap.

KEY FEATURES Red-brown mycelium base.

 DIMENSIONS Cap 1–3cm; stem 4–12 x 0.2–0.8cm.

SPORE PRINT Dark brown-black.

HABITAT AND SEASON Dead deciduous wood; late spring–summer.

EDIBILITY Not edible

Firerug Inkcap
Coprinellus domesticus.

MOTTLEGILLS

Egghead Mottlegill
Panaeolus semiovatus

Mottlegills are so named because of the dark, mottled markings on the gills. This effect is produced as the spores mature unevenly at different rates. Many of these mushrooms are found on dung, and the caps are often small, conical or shallowly convex. The Egghead Mottlegill is always found on dung, and has been appropriately named due to the shape of its smooth, creamy-white cap. There are usually remnants of a small ring left on the tough, whitish stem, which fades to yellowish with age; it sometimes has dark markings at the apex where the maturing black spores have fallen.

KEY FEATURES Egg-shaped cap.

 **DIMENSIONS** Cap 2.5–6cm; stem 4.5–10.5 x 0.4–0.9cm.

SPORE PRINT Black.

HABITAT AND SEASON On dung; spring–late autumn.

EDIBILITY Not edible.

Egghead Mottlegill *Panaeolus semiovatus*.

Turf Mottlegill
Panaeolus fimicola

A particularly urban species, this mottlegill often frequents lawns, especially those with short-cut grass, from as early as March right through to autumn. When young the button-like caps are dark brown, especially when wet. Over time the cap dries to a pale tan colour from the edge inwards, producing several brown shades. The gills are a very light grey-brown colour, changing to mottled black, then completely black. The slender brown stem, which is around 2–5mm thick, is covered in a very fine white, frost-like down, especially at the base.

KEY FEATURES Pale margin when drying out. Frosty stem.

 DIMENSIONS Cap 1.5–4cm; stem 4–8 x 0.3–0.5cm.

SPORE PRINT Black.

HABITAT AND SEASON Short grass, lawns; spring–autumn.

EDIBILITY Not edible. There are similar-looking poisonous species.

Turf Mottlegill
Panaeolus fimicola.

Brown Mottlegill or Brown Hay Cap
Panaeolina foenisecii

This mottlegill always grows in short grass, and is sometimes known as the Mower's Mushroom for this reason. There are similar-looking poisonous brown species. All should be avoided for consumption. The cap is bell-shaped or slightly more convex, and is yellowish to reddish-brown, drying paler from the centre outwards, and creating a two-toned effect similar to that of the Turf Mottlegill (see opposite). The young, pale brown gills have a mottled appearance as the spores mature. The stem is paler than the cap and covered in fine white down at the base.

KEY FEATURES Hygrophanous, often two-toned.

 **DIMENSIONS** Cap 1–2cm; stem 4–7 x 0.2–0.3cm.

SPORE PRINT Brown-black.

HABITAT AND SEASON Short grass; spring–winter.

EDIBILITY Not edible. Poisonous if eaten raw.

Brown Mottlegill *Panaeolina foenisecii*.

OYSTERS

Oyster Mushroom
Pleurotus ostreatus

Oyster mushrooms have distinctive shell-like shapes with little or no stems. Several are edible and choice. *P. ostreatus* is one such mushroom that fruits throughout the year. It grows in clusters on fallen stumps, and standing and fallen trunks, most often on deciduous trees. Apart from the typical shell shape, the white decurrent gills and minimal (or missing) lateral stem are typical characteristics. The cap flattens out as it grows, often becoming wavy or split at the margin. Colours are variable, often occurring in shaded hues of grey/whitish-brown, blue-grey or violet-grey.

KEY FEATURES Shell shape. Minimal stem.

 DIMENSIONS Cap 5–15cm; stem 2–3 x 1–2cm.

SPORE PRINT Whitish-lilac.

HABITAT AND SEASON Usually deciduous stumps, standing and fallen trunks; all year, especially autumn.

EDIBILITY Edible. Good.

Oyster Mushroom *Pleurotus ostreatus*.

Branching Oyster
Pleurotus cornucopiae

This popular cultivated mushroom is grown on dead deciduous tree logs and sold throughout the world. In the wild it is found in tiered clusters on deciduous trees and fallen branches. The cap is typically shell-shaped with an off-centre depression, in line with the stem underneath, which also has very decurrent gills. It is initially white-cream and covered in a whitish bloom. Over time it develops an ochre tint, and eventually becomes completely ochre-brown with a wavy and often split margin.

KEY FEATURES White/ochre shell shape. Decurrent gills.

 **DIMENSIONS** Cap 3–11cm; stem up to 5cm and curved.

SPORE PRINT White.

HABITAT AND SEASON Deciduous tree branches and logs; autumn.

EDIBILITY Edible. Very good.

Branching Oyster
Pleurotus cornucopiae.

Gills are deeply decurrent.

OYSTERLINGS

Variable Oysterling
Crepidotus variabilis

Crepidotus species or oysterlings are much smaller than the popular Oyster Mushrooms (see p. 124), but are similar looking to them. The Variable Oysterling only grows up to 2cm across, but it always occurs in large numbers on woodland litter such as small twigs and dead grass. The small whitish cap is kidney-shaped with an uneven or lobed margin. There is no apparent stem, just a small inconspicuous attachment. The gills are moderately spaced for its size and radiate outwards from the rudimentary stem. They are initially white, and mature to pinkish-brown with age.

KEY FEATURES Small and kidney-shaped. Occurs in large numbers.

 DIMENSIONS Cap 0.4–2cm; stem missing or very minimal.

SPORE PRINT Pinkish.

HABITAT AND SEASON Twigs, straw and dead grass; autumn–winter.

EDIBILITY Not edible. Insubstantial.

Variable Oysterling *Crepidotus variabilis*.

Peeling Oysterling
Crepidotus mollis

The distinctive feature of this common oysterling lies in the fine, skin-like covering on its cap, which is glutinous and stretchy. It can be peeled away if dry enough or not too slimy. The fungi often grow in tiers in large numbers on decaying deciduous tree trunks and stumps. The small, kidney-shaped caps are pale yellow-brown or whitish-brown and have grey-coloured margins. The mushrooms are laterally attached to the wood and there is little or no stem present. The pale gills are very crowded and mature to a light brown colour.

KEY FEATURES Slimy or sticky peeling cap.

DIMENSIONS Cap 1.5–6.5cm; stem missing or very minimal.

SPORE PRINT Brown.

HABITAT AND SEASON Deciduous stumps and logs; late summer–late autumn.

EDIBILITY Not edible.

Peeling Oysterling *Crepidotus mollis*.

SPIKES

Copper Spike
Chroogomphus rutilus (Gomphidius viscidus)

The Copper Spike, like other *Chroogomphus* species, has deeply decurrent and widely spaced gills. Most of the species are edible, but there are mixed opinions on their palatability. As the common name of the Copper Spike suggests, the convex cap is copper-brown in colour with a mild violet hue. It has a slimy texture when wet. The ochre-yellow stem has darker brown markings covering its surface, and the base is bright yellow, as is the inner flesh. The ochre-brown gills become darker brown-purple as the spores mature.

KEY FEATURES Acutely decurrent, widely spaced gills.

 DIMENSIONS Cap 4–14cm; stem 6–11 x 0.3–1cm.

SPORE PRINT
Dusky black to sepia.

HABITAT AND SEASON With conifers, especially Scots Pine; autumn.

EDIBILITY Edible, turns violet when cooked.

Copper Spike
Chroogomphus rutilus.

CHANTERELLES

Chanterelle
Cantharellus cibarius

One of the most popular edible wild mushrooms, the Chanterelle is protected in some areas of Europe so it advisable not to over-pick it. Fortunately it often grows in large numbers. The whole fruiting body is egg-yolk yellow; it is initially flat and matures into a funnel shape with a wavy, inrolled margin. The false decurrent gills are like folds of flesh, often forked. The flesh has a pleasant, mild fruity odour reminiscent of apricots, and a mild peppery flavour. The False Chanterelle (see p. 55) is a deceptive lookalike and should be avoided as some people suffer ill effects after consuming it.

KEY FEATURES Mild fruity odour.

 DIMENSIONS Cap 3–9cm; stem 3–8 x 0.5–1.5cm.

SPORE PRINT Ochre-yellow.

HABITAT AND SEASON Deciduous and coniferous woods; summer–late autumn.

EDIBILITY Edible. Very good.

Chanterelle *Cantharellus cibarius*.

Trumpet Chanterelle *Cantharellus tubaeformis*.

Trumpet Chanterelle
Cantharellus tubaeformis

The Trumpet Chanterelle is edible and pleasant, but less superior in taste to the common Chanterelle (see p. 129). The dark brown cap matures to a funnel shape with a distinctive irregularly wavy margin. The veiny and forked false gills are beige-yellow and have a decurrent attachment. In contrast to the dark cap, the stem is dirty yellow with ridges and flattened areas due to it being hollow inside. The Golden Chanterelle *C. aurora* is a similar species but has wrinkled and waxy gills that are pale yellow-brown.

KEY FEATURES Ochre-yellow stem. Primitive gills.

 DIMENSIONS Cap 2–5cm; stem 5–8 x 0.5–1cm.

SPORE PRINT Yellow.

HABITAT AND SEASON Deciduous and coniferous woods; autumn.

EDIBILITY Edible. Good.

Broadly convex cap with depressed centre.

Horn of Plenty
Craterellus cornucopioides

Often overlooked due its natural camouflage colouring in leaf litter, the Horn of Plenty is a popular edible mushroom. With an intense, powerful taste, it is commonly used for flavouring sauces or dried and used as a seasoning. The whole fruiting body is hollow and tubular-shaped, with an expanding, wavy-edged cavity. There is no definable stem on the young, fragile body. It has brownish-black colouring when moist, and becomes paler and rubbery with age. There are no gills present, just a simple spore-bearing, wrinkly outer surface that is ash-grey in colour.

KEY FEATURES No gills. Trumpet shape.

 DIMENSIONS Cap 2–8cm; stem not applicable.

SPORE PRINT
White.

HABITAT AND SEASON Leaf litter in deciduous woods; late summer–early winter.

EDIBILITY Edible. Strong flavour.

Horn of Plenty
Craterellus cornucopioides.

BOLETES

Cep or Penny Bun
Boletus edulis

Boletes are mushrooms that have pores instead of gills. Densely packed tubes descend beneath the flesh of the cap, their open ends appearing as small 'pores' on the surface underneath. The Cep is a highly prized edible mushroom. Its smooth, convex cap is pale white-ochre when young, gradually turning deep brown. At maturity it can reach up to 30cm across and features a pale white zone around its margin. The sturdy pale stem is thicker at its base than at the apex, and is covered in a fine, raised network of white reticulum. The small pores are white, and turn greenish-yellow with age.

KEY FEATURES White network on stem.

DIMENSIONS Cap 10–30cm; stem 10–20 x 10cm.

SPORE PRINT Olive-brown.

HABITAT AND SEASON Broadleaved and coniferous woods, especially spruce and beech; early summer–autumn.

EDIBILITY Edible. Very good.

Cep or Penny Bun *Boletus edulis*.

Bay Bolete *Boletus badius*.

Bay Bolete
Boletus badius (Xerocomus badius)

A characteristic feature of this chestnut-coloured bolete is that the pores bruise blue on handling. Its broadly convex cap is velvety when young, and becomes smoother with age. In wet or moist conditions it has a slimy, viscous consistency. Mature specimens can grow up to 15cm across, and the relatively long, smooth pale stem has brownish streaky markings. If found growing in locations with high grass, the stem is usually longer and much thinner. The pale lemon-white pores are very small and readily bruise blue-green.

KEY FEATURES Blue bruising.

DIMENSIONS Cap 5–15cm; stem 4–13 x 1–2.5cm.

PORE PRINT Olive-brown.

HABITAT AND SEASON Broadleaved and coniferous woods, especially Scots Pine; late summer–autumn.

EDIBILITY Edible. Very good.

Pores stain blue after bruising.

Suede Bolete or Downy Bolete
Boletus subtomentosus (Xerocomus subtomentosus)

With a velvet-like cap texture that is reflected in its names, this bolete is rounded to broadly convex and red-brown in colour. The surface darkens upon handling and the smooth, pallid stem has a modest network of dark brown veins near the centre. Many boletes feature colour changes to the flesh when cut open and exposed to the air. In this case there is usually a bluish tinge at the top of the stem near the cap and tubes, and ochre-yellow colouring at the base. The large, angular pores bruise blue, but then fade.

KEY FEATURES Velvet cap texture.

 DIMENSIONS Cap 4–10cm; stem 8–10 x 1–1.5cm.

SPORE PRINT Olive-brown.

HABITAT AND SEASON All types of woodland; autumn.

EDIBILITY Edible. Not recommended.

Suede Bolete *Boletus subtomentosus*.

Red Cracking Bolete
Boletus chrysenteron

A small to medium-sized bolete, this species is usually solitary or occurs in small groups. The cap colour ranges from buff-brown to dark red-brown, sometimes with a rosy-red hue. The convex cap is velvety when young and becomes smoother, eventually developing cracks that reveal rosy-red flesh beneath. The lemon-yellow stem is covered in long reddish streaks, and the large yellow pores sometimes discolour greenish upon handling. The exposed flesh turns slightly blue near the top of the stem and above the tubes, and is flushed slightly red further down. Although this mushroom is edible, young, firm specimens are recommended.

KEY FEATURES Cracked cap. Rosy-red colouring.

DIMENSIONS Cap 4–11cm; stem 4–8 x 1–1.5cm.

SPORE PRINT Olive-brown.

HABITAT AND SEASON Broadleaved woods; autumn.

EDIBILITY Edible. Young specimens preferable.

Red Cracking Bolete
Boletus chrysenteron.

Orange Birch Bolete
Leccinum versipelle

Several orange-capped *Leccinum* species are associated with particular trees. This one is always found under birch in open woodland and heathland. The yellow-orange cap initially feels slightly velvety, but soon becomes smooth. The cap cuticle overhangs the rim, often curling inwards. The long whitish stem is peppered with tiny brownish-black scales, and the greyish-white pores turn yellowish with age. Cutting the mushroom in half reveals the whitish flesh within. After a few minutes it changes to violet-black with a bluish hue near the base.

KEY FEATURES Brown-black scales on white stem.

 DIMENSIONS Cap 6–15cm; stem 10–20 x 1.5–4cm.

SPORE PRINT Brown.

HABITAT AND SEASON Under birch in open woodland and heathland; summer–autumn.

EDIBILITY Edible. Good. Must be cooked.

Orange Birch Bolete *Leccinum versipelle*.

Brown Birch Bolete
Leccinum scabrum

The cap colour of the Brown Birch Bolete is variable, ranging from tawny-brown to reddish-brown. The texture is smooth, but more sticky in wet conditions than in dry ones. Longish brown-black scales cover the long white stem, becoming much denser towards the base. The white pores soon discolour to a drab greyish-yellow, bruising yellow-brown on handling. There are no real noticeable colour changes to the exposed inner flesh, except sometimes a very pale pink flush. There is a rarer white variant of this mushroom, but it has a much stockier appearance.

KEY FEATURES Denser scales at stem base than at the top.

DIMENSIONS Cap 7–15cm; stem 8–20 x 2–3.3cm.

PORE PRINT Olive-brown.

HABITAT AND SEASON Under birch; summer–autumn.

EDIBILITY Edible. Must be cooked.

Brown Birch Bolete *Leccinum scabrum*.

Larch Bolete
Suillus grevillei

Suillus species are closely related to the *Boletes*. They mostly feature slimy caps and larger pores, and grow exclusively with conifers. Several species feature rings on their stems. The whole fruiting body of the Larch Bolete is coloured in shades of yellow-orange with a very glutinous and bright yellow (sometimes orange-yellow) cap, which becomes shiny when dry. The yellow, slightly flaky stem is often tinted with orange-yellow and features a pale, skin-like and slimy ring zone. The exposed inner flesh has a yellow hue and is unchanging.

KEY FEATURES Slimy whitish-yellow ring.

 DIMENSIONS Cap 4–11cm; stem 5–9 x 1.3–2cm.

SPORE PRINT
Yellow-brown.

HABITAT AND SEASON With larch; summer–autumn.

EDIBILITY Edible. Remove slimy layer.

Larch Bolete
Suillus grevillei.

Slippery Jack
Suillus luteus

Found exclusively in conifer woodland, especially with Scots Pine, this medium to large *Suillus* species features a distinctively glutinous, sepia-brown cap. The small, rounded, straw-yellow pores can become flushed with a deeper brown colour over time, but initially they are covered with a white veil, which eventually drops to become a large, pendulous ring. It often becomes detached, leaving behind a dark brown-black zone on the white stem. There are usually dark brown streaks below the ring, or ring zone, and paler yellow colouring above.

KEY FEATURES Dark brown, glutinous cap.

DIMENSIONS Cap 6–12cm; stem 5–10 x 1.5–2cm.

PORE PRINT Clay- to ochre-brown.

HABITAT AND SEASON With conifers, especially Scots Pine; autumn.

EDIBILITY Edible. Remove slimy layer.

Slippery Jack *Suillus luteus*.

BRACKET FUNGI

Dryad's Saddle
Polyporus squamosus

Polypores are a group of fungi with pores, also known as bracket or shelf fungi; they grow exclusively on trees or from the ground, feeding off rotting wood underneath. Dryad's Saddle appears from spring through to early autumn, when many blackened, dying specimens can be found fallen on the ground. It grows in densely tiered groups on deciduous trees and stumps. Its semicircular or 'saddle-shaped' cap grows very large, and is yellow-brown and concentrically covered in dark brown fibrous scales. There is a depressed area where it meets the dark and chunky stem. The large, angular pores are creamy-yellow and darken with age.

KEY FEATURES Large, semicircular cap. Dark scales.

 DIMENSIONS Cap 5–60cm; stem 3.5–10 x 1–6cm.

SPORE PRINT White.

HABITAT AND SEASON Deciduous trees and stumps; spring–autumn.

EDIBILITY Edible.

Dryad's Saddle
Polyporus squamosus.

Young Dryad's Saddle.

Hen of the Woods

Grifola frondosa

This large fruiting body is made up of many shell-shaped, wavy caps, grouped together from a singular white branching stem. It is always found at the very bases of broadleaved trees, especially beech and oak, and can reach up to 40cm across. Each grey-brown cap in the floret-like ensemble ranges in size from 4 to 10cm across, and is often wrinkled on the surface. With age the caps become darker as the consistency toughens. The creamy-white pores on the underside continue partway down the stems, becoming larger and elongated.

KEY FEATURES Multiple grey-brown wavy caps.

DIMENSIONS Cap singular 4–10cm; fruiting body up to 40cm; stem not applicable.

PORE PRINT White.

HABITAT AND SEASON Bases of broadleaved trees; autumn.

EDIBILITY Edible. Good when young.

Hen of the Woods *Grifola frondosa*.

Giant Polypore
Meripilus giganteus

Because this polypore is found at the bases of deciduous trees and stumps, growing in a densely grouped fashion, it is sometimes mistaken for the Hen of the Woods (see p. 143), but the fruiting bodies are much larger and often paler. The numerous caps share the same shortly branched stem, ranging from 40 to 90cm across. They are relatively thin in comparison to their width, with a tough but malleable consistency. The light brown colour displays several darker streaked zones, made up of brown fibrous scales. The cap margin is fanned or rosette-like and slightly grooved.

KEY FEATURES Large tiered caps.

 DIMENSIONS Cap singular 10–30cm; fruiting body up to 90cm; stem not applicable.

SPORE PRINT White.

HABITAT AND SEASON Bases of deciduous trees; autumn.

EDIBILITY Edible. Slow cook.

Giant Polypore
Meripilus giganteus.

Chicken of the Woods or Sulphur Polypore
Laetiporus sulphureus (Polyporus sulphureus)

This is an attractive bracket fungus that fruits early in the season. The large, semicircular or fanned brackets grow in tiered formations on the sides of deciduous trees, reaching up to 40cm across. They are irregularly wavy or lumpy and feature striking lemon-orange colouring with small lemon-yellow pores on the underside. The texture is smooth and suede-like. The bright colouring gradually fades with age. Young layered brackets are thick and fleshy, and this is the best time to collect them for eating. Older and tougher specimens are found during late summer to autumn and are not edible.

KEY FEATURES Sulphur-yellow pores.

DIMENSIONS Individual brackets 10–40cm across.

SPORE PRINT White.

HABITAT AND SEASON On deciduous trees, especially oak; spring–autumn.

EDIBILITY Edible. Good. Slightly bitter.

Chicken of the Woods *Laetiporus sulphureus*.

Beefsteak Fungus
Fistulina hepatica

Appearing as a small, pinkish-red, cushion-like ball when very young, this distinctive bracket fungus expands into a semicircular or tongue-like shape up to 30cm across, with little or no stem. The whole fruiting body is thick, and pink-red to brown-red, with flesh that has the consistency of raw meat. Squeezing a young specimen will cause it to 'bleed' a red juice. The species is a parasite on living oak trees, often growing at the bases. It is edible but must be soaked in milk or water to remove the bitter tannic acid.

KEY FEATURES 'Meaty' flesh. 'Bleeding' juice.

 DIMENSIONS Bracket 10–30cm across; 2.5–6.5cm thick.

SPORE PRINT White or pale ochre.

HABITAT AND SEASON On oak; late summer–autumn.

EDIBILITY Edible. Can be bitter.

Beefsteak Fungus
Fistulina hepatica.

Southern Bracket
Ganoderma australe

Very similar in appearance to Artist's Bracket *G. applanatum*, the Southern Bracket features a much thicker, rounded white margin during the growing season. The dark brown, semicircular bracket can grow to up to 40cm across, and is concentrically ridged and lumpy with a hard outer crust. The distinctive feature of this bracket, as well as of Artist's Bracket, is that the whitish pores on the underside readily bruise dark brown. If marked with a fingernail or twig, the contrasting colour provides a perfect medium to write or draw on.

KEY FEATURES Immediate dark brown change on marked pores.

DIMENSIONS Bracket 5–40cm across; 5–30cm thick.

SPORE PRINT Dark brown.

HABITAT AND SEASON Deciduous trees; all year, persisting.

EDIBILITY Not edible. Too tough to eat.

Southern Bracket
Ganoderma australe.

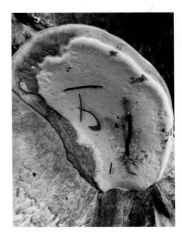

Pores bruise dark brown once marked.

Birch Polypore or Razorstrop Fungus
Piptoporus betulinus

This is an extremely common polypore fungus that is found
exclusively on standing or fallen birch trees throughout the year.
Typically semicircular or kidney-shaped, it can grow to up to 25cm
across by 7cm thick. Initially pure white, it matures to a dull
greyish-white or even tawny-brown colour. The smooth surface
often cracks, exposing the white flesh underneath. The
consistency is spongy or slightly rubbery. The pure white pores
are smooth, later becoming greyish. In the past this fungus was
valued for its antibacterial properties and ability to sharpen
blades or tools, hence the name Razorstrop Fungus.

KEY FEATURES Kidney-shaped. Off-white colouring.

 DIMENSIONS Bracket 25cm across; 2.5–7cm thick.

SPORE PRINT White.

HABITAT AND SEASON On living or dead birch; all year.

EDIBILITY Not edible. Tough and bitter.

Birch Polypore *Piptoporus betulinus*.

Hoof Fungus or Tinder Bracket
Fomes fomentarius

This perennial bracket fungus has a very hard, wood-like outer crust. Each year it produces newly grown layers, sporulating through spring and summer. After some time it develops a distinctive 'hoof' shape with tiered, concentric, grey-banded ridges. Several fruiting bodies can often be found on the same tree. The minute pores are initially pale grey-brown, and eventually turn dark brown. The leathery, soft brown inner flesh can be used as tinder or as a leather substitute which has been used for making hats and bags. The hollowed out fruiting body was also once used for keeping and transporting burning embers.

KEY FEATURES Hard, hoof shape.

DIMENSIONS Bracket 5–40cm across; 5–20cm broad; 8–20cm deep.

SPORE PRINT White.

HABITAT AND SEASON On dead or dying birch and beech; all year.

EDIBILITY Not edible.

Hoof Fungus *Fomes fomentarius*.

Blushing Bracket
Daedaleopsis confragosa

The semicircular, fan-shaped brackets of this species usually grow in tiered groups. The reddish-brown upper surface shows distinctive concentrically radiating 'bands' with a wrinkled texture (smoother when young). The bracket thickens at the point of attachment to the wood, and the margin remains thin and undulating. Young specimens are white, sometimes with a flush of pink. The large, whitish pores are a mix of elongated slots and round holes, and readily bruise pinkish-red after rubbing. This feature is unique to this bracket fungus. With age the fruiting body turns dark red-brown and becomes tough and corky.

KEY FEATURES Pink-red bruising on underside.

 DIMENSIONS Bracket 5–15cm across; 4–10cm broad; 1–4cm thick.

SPORE PRINT
White.

HABITAT AND SEASON On deciduous trees; all year.

EDIBILITY Not edible.

Blushing Bracket
Daedaleopsis confragosa.

Hairy Curtain Crust or Hairy Stereum

Stereum hirsutum

The Hairy Curtain Crust grows in layers on the dead or fallen wood and stumps of deciduous trees. It can be found all year round. The zoned yellow-brown fruiting bodies are semicircular and wavy, typically forming in several rows and often overlapping each other. The Latin word *hirsutum* describes the fine 'hairy' tufts on the upper surface of the fungus. The brighter yellow-orange underside is smooth, and ages to a dull grey-brown. Older groups are often green in colour due to a fine layer of algae covering the surface.

KEY FEATURES Fine hairy fibres on upper side.

DIMENSIONS Bracket 2–6cm across; 0.2–0.3cm thick.

SPORE PRINT White.

HABITAT AND SEASON Dead deciduous wood; all year.

EDIBILITY Not edible. Too tough to eat.

Hairy Curtain Crust *Stereum hirsutum*.

Turkeytail
Trametes versicolor

Often occurring in large, tiered groups on deciduous wood, this attractive fungus can be found all year round. It has a very distinctive fan-like shape with concentric bands; the colours are often variable, in combinations such as browns, blues, greys and greens. The thin and wavy margin is reliably coloured, always remaining creamy-white. When young the upper surface has a soft and velvety texture; this becomes much smoother with age. The tiny creamy-white pores are round with a few angular exceptions. The tough, leathery white flesh is thin with no real taste. The species is not recommended for eating, but it is sometimes used as a decoration in flower arrangements.

KEY FEATURES Zoned colours. White margin.

 DIMENSIONS Bracket 4–10cm across; 0.3–0.5cm thick.

SPORE PRINT White.

HABITAT AND SEASON Dead deciduous wood; all year.

EDIBILITY Not edible.

Turkeytail *Trametes versicolor*.

Colour combinations vary.

TOOTHED FUNGI

Wood Hedgehog or Hedgehog Mushroom
Hydnum repandum

A small group of mushrooms bear spines, also known as teeth, in place of gills or pores, where their spores develop and eventually disperse. The Wood Hedgehog is one of the most common of these fungi and is particularly prized as an edible species. It has a convex creamy-yellow to pale ochre cap, and the chunky white stem sometimes has an off-centre attachment, producing a slight depression in the cap. It is smooth and very suede-like at maturity. The white or sometimes pinkish spines can reach up to 6mm long. Old specimens drop their spines when handled or knocked.

KEY FEATURES Creamy-yellow. Spines.

 DIMENSIONS Cap 4–18cm; stem 4–8 x 1.5–4cm.

SPORE PRINT White.

HABITAT AND SEASON Deciduous and coniferous woods; late summer–late autumn.

EDIBILITY Edible. Young specimens very good.

Wood Hedgehog *Hydnum repandum*.

PUFFBALLS

Giant Puffball
Calvatia gigantea

Puffballs are sack fungi, maturing their spore mass, known as a gleba, within a ball-like organic case. The Giant Puffball is unmistakable due to its very large size – it grows up to 90cm in diameter, or even larger in rare cases. The white spore mass is protected by the leathery white outer casing (edible at this stage), and matures to olive-brown. The outer skin gradually peels away and the fruiting body becomes uprooted from its weak mycelial cord, at which point it is free to disperse trillions of spores, of which only a couple successfully reproduce. Old specimens turn completely brown.

KEY FEATURES Very large ball shape.

 DIMENSIONS Fruiting body 10–90cm in diameter.

GLEBA White, then olive-brown.

HABITAT AND SEASON Gardens, pastures and woods; summer–autumn.

EDIBILITY Edible when spores still white.

Giant Puffball *Calvatia gigantea*.

Common Puffball
Lycoperdon perlatum

Often found in small, scattered groups in open woodland, the young fruiting body is pure white, and is covered in many pyramid-shaped nodules on the surface that are usually denser at the top. It has a noticeable lower stem shape that is often quite long and is loosely attached to the soil by a mycelial cord. With age the colour becomes dull yellow-brown and a small cavity opens at the apex to release the mature olive-brown spores. Raindrops, wind or movement from passing animals cause the open sack to 'puff' out its contents in a fine cloud of brown powder.

KEY FEATURES Warty nodules on outer skin.

DIMENSIONS Fruiting body 2–6cm across; 3–9cm high.

GLEBA White, then olive-brown.

HABITAT AND SEASON Woods; late summer–autumn.

EDIBILITY Edible when spores still white.

Common Puffball
Lycoperdon perlatum.

Open cavity, releasing spores.

Stump Puffball *Lycoperdon pyriforme*.

Stump Puffball
Lycoperdon pyriforme

The Stump Puffball is easy to identify as it is the only puffball that grows on wood, often in huge numbers covering rotting stumps and logs. The young whitish-yellow fruiting bodies are often club-shaped and covered in fine granule-like warts. With age their colour becomes dull brown, sometimes with a grey hue. The mature olive-brown spores are released via a cavity at the apex. There are noticeable white mycelial threads attaching the base of the puffball to the substrate. Although edible when young, this species has an unpleasant taste.

KEY FEATURES Grows on wood.

DIMENSIONS Fruiting body 1.5–3.5cm across; 3–5cm high.

GLEBA White, then olive-brown.

HABITAT AND SEASON Stumps and logs; summer–autumn.

EDIBILITY Edible. Not recommended.

Older examples with open cavities to release the spores.

Meadow Puffball
Vascellum pratense

As its common name suggests, this puffball is often found in meadows and other grassy locations such as lawns, pastures and even golf courses. The young fruiting bodies are creamy-white, irregularly rounded and often squat with a short, thick, stem-like base. The entire surface area is covered with fine, powdery spines that brush away easily when touched. As the pure white spores mature to olive-brown, the outer skin turns to a dull light brown and develops a spore-releasing cavity near the apex. The cavity expands over time and older specimens appear like small bowls.

KEY FEATURES Powdery white surface.

 DIMENSIONS Fruiting body 2–5cm across; 2.5–4.5cm high.

GLEBA White, then olive-brown.

HABITAT AND SEASON Grassland; summer–autumn.

EDIBILITY Edible when spores still white.

Meadow Puffball *Vascellum pratense*.

EARTHBALLS

Common Earthball
Scleroderma citrinum

The Common Earthball often occurs in small scattered groups throughout the summer and autumn months, in rich peaty soil or moss. This mycorrhizal species shares a special relationship with deciduous trees, especially oak, beech and birch. The fruiting body is dirty yellow to ochre-brown with an irregular ovoid or potato-like shape, and is covered with rough, dark brown scales. When young, it is solid with a thick outer-wall casing protecting the purple-black spore mass inside. At maturity the gleba turns into a fine powder and the outer surface splits in random patches to release the spores. White, cotton-like mycelial cords attach the fruiting body to the soil.

KEY FEATURES Thick outer wall.

DIMENSIONS Fruiting body 2.5–10.5cm across.

GLEBA Purple-black.

HABITAT AND SEASON Damp woodland; summer–autumn.

EDIBILITY Poisonous.

Common Earthball
Scleroderma citrinum.

'Potato-like' shape.

Scaly Earthball
Scleroderma verrucosum

The Scaly Earthball features a thick, long tapering stem, which is often irregularly shaped or ribbed and pale yellow-white in colour. It is covered in white mycelial cords nearer the base. The main yellow-brown fruiting body is ellipsoid in shape, and often flatter at the apex. It is spotted with smooth, small dark brown scales. The outer skin is very thin, especially when compared to that of the Common Earthball (see p. 159). A gentle squeeze easily deforms its shape. At maturity the outer skin irregularly splits near the apex to release the powdery dark brown spores.

KEY FEATURES Long mycelial stem.

 DIMENSIONS Fruiting body 3–5.5cm across.

GLEBA Olive-brown

HABITAT AND SEASON Woodland, in soil; summer–autumn.

EDIBILITY Poisonous.

Scaly Earthball
Scleroderma verrucosum.

Leopard Earthball
Scleroderma areolatum

One of the smaller earthballs, the Leopard Earthball is rounded or ellipsoid in shape with a tapered rooting stem. Fine mycelial strands are attached at the very bottom. The pale yellow-brown outer wall is thin with a smooth texture and features a very distinctive scaly pattern. Small, dark and rounded scales, each one encircled by a dark ring, cover the surface. With age the rings fade, leaving a net-like pattern on the surface. The inner spore mass, or gleba, is purple-brown. When mature the outer casing splits open to free its spores.

KEY FEATURES Small, dark, ringed scales.

 DIMENSIONS Fruiting body 1.5–4.5cm across.

GLEBA Purple-brown.

HABITAT AND SEASON Clearings in soil and moss; late summer–autumn.

EDIBILITY Poisonous.

Leopard Earthball *Scleroderma areolatum*.

EARTHSTARS

Collared Earthstar
Geastrum triplex

Earthstars are stomach fungi. They are unique in appearance because of the star shape produced when a mature specimen's outer layer opens up into a petal-like shape. The Collared Earthstar is one of the most common species. The round fruiting body or spore sac is initially encased in a thick, fleshy outer layer and is distinctly bulb-shaped. The outer layer peels away, splitting into several petal-like segments. The segments continue to curl back on themselves, often cracking or splitting. The spore sac develops a tear at the apex, ready to release the spores into the wind.

KEY FEATURES 4–8 'petal' shapes.

 DIMENSIONS Fruiting body 3–6cm across when closed; up to 11cm across when open.

SPORE MASS Dark brown.

HABITAT AND SEASON Deciduous woods; summer–autumn.

EDIBILITY Not edible.

Collared Earthstar
Geastrum triplex.

Spore sac exposed.

BIRD'S NEST FUNGI

Shooting Star
Sphaerobolus stellatus

Bird's nest fungi are very small, featuring nest-like fruiting bodies that contain small, egg-like spore sacs known as peridioles. The Shooting Star is a very small species, only around 2mm across, found on all types of woodland debris, where it is easy to miss. The rounded fruiting body is whitish-yellow, maturing darker as it splits open from the apex into several petal-like segments, and creating a distinctive star shape. The exposed peridioles are brown and spherical. When ready, they eject the spores several metres up into the air, hence the species' common name.

KEY FEATURES Very small with collared, star-like segments.

DIMENSIONS Fruiting body 0.15–0.3cm across.

PORES White-yellow.

HABITAT AND SEASON Woodland debris, dung, sawdust and so on; autumn.

EDIBILITY Not edible.

Shooting Star *Sphaerobolus stellatus*.

Common Bird's Nest
Crucibulum laeve

This common species is often found in large numbers on woodland debris such as twigs, bark litter and coniferous tree cones. Initially the fruiting body is cylindrically drum-shaped, with white sides and a scurfy ochre-yellow upper surface that acts as a protective layer, covering the immature spore sacs. It soon develops into a distinctive cup shape with an ochre-brown outer shell. The upper sheath slowly breaks apart, exposing the seed-like peridioles inside. These are attached to the 'nest' with a thin cord (funiculus). The inner wall of the fruiting body is off-white to greyish, smooth and shiny.

KEY FEATURES Brown cup shape.

 DIMENSIONS Fruiting body 0.5–1cm across; 0.3–0.8cm high.

SPORES White-yellow.

HABITAT AND SEASON On woodland debris; autumn–early spring.

EDIBILITY Not edible.

Common Bird's Nest
Crucibulum laeve.

Early 'drum-shaped' stage.

STINKHORNS

Stinkhorn or Witch's Egg

Phallus impudicus

The Stinkhorn initially appears as a large, ball-shaped egg, half buried in the soil. It soon ruptures at the apex to expose a gelatinous layer under the skin. A fragile white, stalk-like growth extends out of the sac, featuring a bell-shaped head covered in dark olive-green slime; this contains the spores. It smells intensely of rotting meat and attracts flies, which in turn spread the spores. Eventually the head is cleared of all slime, leaving a white honeycomb surface pattern that is deeply ribbed.

KEY FEATURES Foul odour. Phallic shape.

DIMENSIONS Egg 3–6cm; stalk 10–20 x 2–3cm.

PORES Pale yellow.

HABITAT AND SEASON Woods, in leaf litter, and gardens; summer–autumn.

EDIBILITY Nut-like core is edible in egg form.

Stinkhorn
Phallus impudicus.

Honeycomb pattern on
exposed head.

Dog Stinkhorn
Mutinus caninus

Although it is similar to the Stinkhorn (see p. 165), the Dog Stinkhorn has a much smaller egg form and stalk. Initially the Dog Stinkhorn is housed in a small, fleshy ball that is half buried in the soil. The ball soon ruptures at the very top. The long, whitish-yellow stalk extends upwards out of the sac. It has a subtly pitted surface texture and is often slightly curved. The apex is covered in very dark olive slime, often exposing an orange-red tip. The pungent smell it produces attracts flies, which in turn spread the spores it contains, leaving a bare, orange-red honeycomb tip.

KEY FEATURES Orange-red tip.

 DIMENSIONS Egg 1–3cm; stalk 4–8 x 0.8–1cm.

SPORES Pale yellow.

HABITAT AND SEASON Woods, in leaf litter, and rotting wood; summer–autumn.

EDIBILITY Not edible.

Dog Stinkhorn
Mutinus caninus.

CLUB FUNGI

Apricot Club
Clavulinopsis luteoalba

Nearly all club fungi are fleshy, spindle-shaped with a rubbery texture and appear in a variety of striking colours. The Apricot Club is wavy, relatively thick and singular, with no branches or multiple forking tips. It is dark yellow or apricot in colour, often with a paler (or whitish) tip, and grows in small clumped tufts (or solitarily) scattered around in short grass. The spores mature on the surface and are spread by the wind and rain. The Yellow Club *C. helvola* is a very similar-looking species but is brighter yellow, and is more often found in open woodland as well as in short grass and moss.

KEY FEATURES Deep yellow or apricot colour. Singular tip.

DIMENSIONS 3–6cm tall; 0.1–0.5cm thick.

SPORES White.

HABITAT AND SEASON Short grassland, lawns; autumn.

EDIBILITY Not edible.

Apricot Club *Clavulinopsis luteoalba*.

CAULIFLOWER FUNGI

Cauliflower Fungus or Wood Cauliflower
Sparassis crispa

The Cauliflower Fungus is a popular edible fungus that is extremely easy to identify due to its unique appearance. The large but fragile fruiting body is rounded and somewhat compressed. It consists of many branchlets, similar to a common cauliflower, that are wavy or lobed at the rim. The entire fruiting body is whitish-yellow or sometimes slightly darker. It is found exclusively with coniferous trees, most often pine. *S. laminosa* is a very similar species, often growing on beech and oak; it is equally edible, but rarer.

KEY FEATURES Large fruiting body. Crisped branchlets.

 DIMENSIONS Fruiting body 20–60cm across.

SPORES Pale yellow-brown.

HABITAT AND SEASON With coniferous trees, especially pine; late summer–autumn.

EDIBILITY Edible. Very good when young.

Cauliflower Fungus *Sparassis crispa*.

CORAL FUNGI

Grey Coral
Clavulina cinerea

Coral fungi have a tight 'branched' appearance similar to that of real coral. The Grey Coral is small and appears in dense tufts that are usually found scattered around on the woodland floor. The ash-grey fruiting body is made up of several branched arms that originate from one (or sometimes more) common base. The tips subdivide again to form a multi-crested apex. The White Coral *C. coralloides* is almost identical in appearance, but is white to off-white in colour.

KEY FEATURES Crested tips. Smoky-grey.

DIMENSIONS Fruiting body 3–11cm tall.

PORES White.

HABITAT AND SEASON All types of woodland on soil; summer–autumn.

EDIBILITY Edible. Bland.

Grey Coral *Clavulina cinerea*.

JELLY FUNGI AND SIMILAR SPECIES

Jelly Ear
Auricularia auricula-judae

The Jelly Ear is extremely popular in Chinese cooking. Even though it is quite flavourless, its crunchy consistency is ideal for dishes such as stir-fries and soups. When young it is thickly fleshed, almost cup-shaped and extremely soft, with a velvet-like texture. As it grows it becomes smoother and irregularly lobed or deformed, appearing very much like a human ear, especially due to its fleshy pinky-brown colouring. It is often found in large numbers on the branches of deciduous trees, especially elder.

KEY FEATURES Ear shape. Pinkish flesh.

 DIMENSIONS Fruiting body 2–8cm across.

SPORES White.

HABITAT AND SEASON On deciduous tree branches and trunks all year, but especially autumn and winter.

EDIBILITY Edible. Minimal taste, crunchy consistency.

Jelly Ear *Auricularia auricula-judae*.

The wrinkled 'ear-like' appearance of Jelly Ear.

Witches' Butter
Exidia glandulosa

This widespread and common jelly fungus is found on the dead wood of deciduous trees. The fungi are shiny black and gelatinous, with tiny warts dotted randomly on the smooth black surface. The fruiting bodies often overlap each other. They are attached to the wood by very tiny stems that are only noticeable once they have been removed from the wood. The inner jelly-like flesh is brown and semi-translucent. After wet weather the fungi become quite conspicuous, full and fleshy. In prolonged dry weather they shrivel up to hard membranous lumps; they soon rehydrate after rain, so can survive all year round.

KEY FEATURES Warty dots on surface.

 DIMENSIONS Fruiting body up to 5cm across.

SPORES White.

HABITAT AND SEASON Dead deciduous tree wood; autumn–winter.

EDIBILITY Not edible.

Witches' Butter *Exidia glandulosa*.

Common Jellyspot
Dacrymyces stillatus

The Common Jellyspot's spherically compressed fruiting bodies are very small, no more than a few millimetres across, but they appear in large numbers on all types of damp, decaying wood throughout the year; they are often found on old fences and wooden gates and posts. Larger blob-like clumps often appear, but these are actually several fruiting bodies that have merged together. The fungi are usually orange-yellow, but there is a paler yellow variant. They dry out with age, becoming darker and slightly tougher.

KEY FEATURES Jelly-like. Very small.

 DIMENSIONS Fruiting body 0.2–0.5cm across.

SPORES White.

HABITAT AND SEASON Dead broadleaved and coniferous wood; all year.

EDIBILITY Not edible.

Common Jellyspot
Dacrymyces stillatus.

Yellow Stagshorn
Calocera viscosa

The Yellow Stagshorn always grows on dead and rotting coniferous tree stumps and logs; it sometimes appears to be growing from the ground, but in these cases it is growing on rotting wood under the soil. The fruiting body is yellow to deep orange-yellow and has a rubbery texture. It features several branchlets that often split in two, each with branched tips. With age it becomes darker and develops a tough consistency. The Small Stagshorn *C. cornea* is very similar but much smaller. It grows on deciduous wood and has no forking branchlets, but instead tapers out to a single thin tip.

KEY FEATURES Branched tips. Only on coniferous wood.

 DIMENSIONS 3–9cm tall.

SPORES White.

HABITAT AND SEASON Conifer stumps and logs; late summer– late autumn.

EDIBILITY Not edible. Sometime used as a garnish.

Yellow Stagshorn *Calocera viscosa*.

MORELS

Common Morel
Morchella esculenta

The Common Morel is a popular and much sought-after edible mushroom. It can be cooked fresh or dried for future use. The cap is variable in shape, ranging from ovoid to irregularly conical or elongated. It is pale yellow-brown or grey-brown, becoming much darker with age. The cap is covered in deep angular pits with raised ridges, forming a non-uniform, honeycomb-like pattern. The stem is yellow-white, swollen at the furrowed base and has a minutely granular texture. The entire fruiting body is hollow.

KEY FEATURES Irregular honeycomb shape.

 DIMENSIONS Cap 6–18cm tall x 3–8cm wide; stem 3–8 x 1.5–3.5cm.

SPORES White to cream.

HABITAT AND SEASON Under broadleaved trees, open woodland and scrub; April–May.

EDIBILITY Edible. Very good. Must be cooked.

Common Morel
Morchella esculenta.

Black Morel
Morchella elata

The Black Morel is very similar to the Common Morel (see p. 175), but has a darker, thinner and more conically shaped cap. It has become more common in urban locations as its spores and mycelia have been transported in wood chippings that have been scattered over gardens. The exterior honeycombed pattern is a reliable identification feature as the deep pits uniformly flow upwards in parallel lines. The pitted flesh is dull olive-brown and the fine exterior ridges are black. The white stem is cylindrical, often irregularly shaped with a granular texture and attached completely to the cap. The entire fruiting body is hollow.

KEY FEATURES Honeycomb pattern, ribbed.

DIMENSIONS Cap 6–15cm tall; stem 3.5–7 x 2–3.5cm.

SPORES White to cream.

HABITAT AND SEASON Coniferous woods, woodchips in gardens; March–June.

EDIBILITY Edible. Good. Must be cooked.

Left: Black Morel *Morchella elata*.

Right: Cap conical, with honeycomb-like surface pattern.

Semifree Morel
Morchella semilibera

The Semifree Morel has a much smaller cap than other morels and is distinctly more conical; it is rarely ovate, and has vertically flowing ochre or dark brown honeycombed pits. The raised, irregularly ridged network is dark brown, turning more blackish with age. The white stem is very long in comparison to the smaller cap. It is only partially joined, or semi-free, at the point where it meets the cap, whereas other common morels are fully attached. The surface is covered in fine granules, giving it a mild scurfy texture.

KEY FEATURES Small conical cap, partly free from stem.

 DIMENSIONS Cap 2.5–6cm tall; stem 3.5–8.5 x 1.5–2.5cm.

SPORES White to cream.

HABITAT AND SEASON All types of woodland; March–June.

EDIBILITY Edible. Can cause stomach upsets.

Semifree Morel
Morchella semilibera.

Cap is semi-free from stem.

False Morel
Gyromitra esculenta

The False Morel is a very poisonous mushroom that can cause liver damage even after being cooked. The poison can be filtered out with repeat cooking and drying, but there is no guarantee that the toxins will be fully removed. It is therefore best avoided. The fruiting body resembles those of many morels, but has no honeycomb pattern. Instead, the fleshy cap is twisted and lobed – it is very brain-like in appearance with an overall deep red-brown colour. The short, creamy-white stem is often deformed and ridged. The whole fruiting body is hollow with a chambered inner structure.

KEY FEATURES Brain-shaped cap.

 DIMENSIONS Cap 3–10cm across; stem 2–4 x 0.2–0.5cm.

SPORES Cream.

HABITAT AND SEASON With conifers, especially pine; April–June.

EDIBILITY Poisonous. Can be fatal.

False Morel
Gyromitra esculenta.

Right: Cap is 'lobed' and twisted.

SADDLES

White Saddle
Helvella crispa

The cap of the White Saddle is naturally white or creamy-white, and has a wax-like texture and is often distorted, with a deep central depression where it meets the stem. With age the flesh becomes lobed and wavy with an overhanging rim. The underside surface is slightly darker yellow-brown, and the hollow stem is white and deeply furrowed. The Elfin Saddle *H. lacunosa* is in many ways identical to the White Saddle, but has a contrasting grey-black colour.

KEY FEATURES Cap saddle-shaped.

 DIMENSIONS Cap 2–6cm tall; stem 2–7 x 1–2.5cm.

SPORES White or creamy-yellow.

HABITAT AND SEASON All types of woodland, paths and lawns; summer–autumn.

EDIBILITY Edible. Must be cooked well.

White Saddle *Helvella crispa*.

CUP FUNGI

Bleach Cup
Disciotis venosa

Many ascomycetes are deeply concave or cup-shaped, so they are naturally referred to as cup fungi. The Bleach Cup is closely related to morels (see p. 175–179) and is often found in proximity to them. The fruiting body is deeply cup-shaped, often with an irregular or split rim. Its whitish outer surface is covered in small, dark brown scales, giving it a scurfy texture. The inner surface is dark chocolate-brown, and smoothly textured with many folds, creases and bumps. The short, thick stalk attachment is buried in the soil. The flesh has a chemical smell, similar to that of bleach or disinfectant.

KEY FEATURES Bleach-like chemical smell.

DIMENSIONS Fruiting body 4–16cm across.

SPORES Pale yellow.

HABITAT AND SEASON All types of woodland; April–June.

EDIBILITY Poisonous.

Bleach Cup
Disciotis venosa.

JELLY DISCS

Purple Jelly Disc
Ascocoryne sarcoides

The Purple Jelly Disc is relatively common and is often overlooked due to its small size, although fruiting bodies can be found in large (often overlapping) groups, when they become more conspicuous. Depending on age, the average jelly-like fruiting body is cup-shaped or has a shallow central depression. The fruiting bodies appear to be attached directly to the wood, but often have short and stout stem attachments. The colour varies from deep pink to darker purple, often with an irregularly shaped or wavy margin.

KEY FEATURES Purple. Disc-shaped.

 DIMENSIONS Fruiting body 0.2–1.7cm across.

SPORES White or creamy-yellow.

HABITAT AND SEASON Dead wood, favouring beech; summer–late autumn.

EDIBILITY Not edible.

Purple Jelly Disc *Ascocoryne sarcoides*.

Lemon Disco
Bisporella citrina

Each individual fruiting body is saucer-shaped and very small, no more than 2–3mm across (often smaller), but this fungus grows in very large colonies that cover the surfaces of dead deciduous wood such as large fallen branches or trunks. The species has a distinctive bright yellow colouring and becomes very conspicuous as the slimy fruiting bodies often overlap to merge with each other. There is a very tiny (or sometimes absent) stem attachment. The Common Jellyspot (see p. 173) grows in similar large groups, but is not as brightly coloured or saucer-shaped.

KEY FEATURES Very small, saucer-shaped.

DIMENSIONS Fruiting body 0.1–0.3cm across.

SPORES Creamy-white or yellow-white.

HABITAT AND SEASON On dead deciduous wood; autumn.

EDIBILITY Not edible.

Lemon Disco *Bisporella citrina*.

XYLARIACEAE

Dead Man's Fingers
Xylaria polymorpha

This peculiar fungus earned its common name due to its uncanny resemblance to a blackened finger rising out of the ground. Several may grow together side by side, further enhancing the strange effect. The entire fruiting body is black with a tough, scurfy crust. It is usually club-shaped but can also be cylindrical or irregularly bulbous. A thin, tapering stem, which can be easily broken, attaches it to the substrate. The inner cavity contains the spore-producing flesh, which is very tough and pure white. Dead Moll's Fingers *X. longipes* is a similar species, but smaller and noticeably thinner.

KEY FEATURES Club- or finger-shaped.

 DIMENSIONS Fruiting body 3–9cm tall; 1–3cm across.

SPORES White.

HABITAT AND SEASON Dead wood and stumps of broadleaved trees; all year.

EDIBILITY Not edible.

Dead Man's Fingers
Xylaria polymorpha.

Candlesnuff Fungus or Stag's Horn
Xylaria hypoxylon

This small fungus is an extremely common species that is present all year round, often in large groups, growing on all types of rotting wood and stumps. The flattened and distorted fruiting body has a distinctive branched appearance, reminiscent of stag antlers. Each branchlet is covered in a fine white powder that turns black at the tip as it matures. Eventually the powder completely dissipates as the fungus dies. The lower black stem has a slightly hairy, scurfy texture. The Beechmast Candlesnuff *X. carpophila* is very similar but much thinner, has fewer branchlets and grows exclusively on rotting beechmast.

KEY FEATURES White powder-covered branchlets.

DIMENSIONS Fruiting body 2–6cm tall.

SPORES White or cream.

HABITAT AND SEASON Dead wood and stumps; all year.

EDIBILITY Not edible.

andlesnuff Fungus *Xylaria hypoxylon*.

King Alfred's Cakes or Cramp Balls
Daldinia concentrica

The tough, blackened fruiting bodies of this species grow on the dead wood of broadleaved trees, mainly ash and beech. It is a very distinctive-looking fungus, similar in appearance to lumps of smooth charcoal. The young fruiting bodies are red-brown in colour with a dull surface, while mature specimens are black with a smooth, shiny surface. The species has a hemispherical shape, and the inner crusty-textured flesh is concentrically patterned with contrasting black-and-white zones (hence *concentrica*). The flesh can be used as tinder, and has also been utilised as a folk remedy for night cramps. Its season is summer to late autumn, although older specimens remain for several years.

KEY FEATURES Black hard lump shapes.

 DIMENSIONS Fruiting body 3–11cm across.

SPORES Black.

HABITAT AND SEASON Dead wood; all year.

EDIBILITY Not edible.

King Alfred's Cakes *Daldinia concentrica*.

TRUFFLES

Summer Truffle
Tuber aestivum

There are many choice truffles located throughout Europe, with some of the best occurring only in Italy and France. The very good and popular Summer Truffle is one of the few truffles to be found in nearly all European countries. The irregularly rounded, dark brown (sometimes blackish) fruiting body is covered with small, pyramid-shaped warts. The inner flesh is initially pure white, and matures to grey-brown or yellow-brown with a distinctive marbled appearance. Like all truffles, Summer Truffles grow buried underneath the soil, most often near beech trees and sometimes oak trees.

KEY FEATURES Pyramid-shaped warts.

DIMENSIONS Fruiting body 3–8cm across.

INNER FLESH White, then grey-brown, marbled.

HABITAT AND SEASON Buried, mainly under beech; summer–autumn.

EDIBILITY Edible. Very good. Pleasant nutty taste.

Summer Truffle *Tuber aestivum*.

GLOSSARY

annulus: ring on the stem.

cap: fleshy attachment on top of the stem, with gills, pores or spines on the underside.

deliquesce: to become liquid (as, for example, in inkcaps).

central boss: see umbo.

fruiting body: the actual mushroom or body of the fungus, grown from the mycelium to fruit and reproduce (disperse spores).

gleba: fleshy, spore-bearing inner mass of certain fungi such as puffballs and earthballs.

gills: spore-depositing, blade-like flesh under the cap of a mushroom, radiating out from the stem.

hyphae: very fine, hair-like filaments, which branch out to form a mycelium.

margin or rim: edge of cap.

mushroom or toadstool: non-scientific name for fungi with caps and stems.

mycelium (plural mycelia): cobweb-like network of hyphae; the fungal organism.

mycorrhizal association: beneficial relationship between fungi and plants; in exchange for sugars, the fungus supplies moisture and nutrients to the plant.

partial veil: see veil.

pores: small, spore-depositing holes at the ends of hollow tubes on the cap undersides of certain mushrooms, such as boletes.

rim: see margin.

spines: spore-bearing spines on the underside of the cap (for example in Hedgehog Mushroom).

spores: minute asexual reproductive units.

spore print: coloured print from fallen spores, collected on a card from the gills, pores or spines of a fungus.

toadstool: see mushroom.

umbo: raised central bump on cap.

universal veil: membranous tissue that fully envelops the young fruiting bodies of certain mushrooms.

veil or partial veil: temporary membranous tissue that covers the gills.

volva: membranous cup-like casing at the base of the stem; remnant of the universal veil.

ORGANISATIONS

The Association of British Fungus
Groups
Harveys, Alston
Axminster, Devon
EX13 7LG
www.abfg.org

British Mycological Society
Wolfson Wing Jodrell Laboratory
Royal Botanic Gardens
Kew, Richmond
Surrey TW9 3AB
www.britmycolsoc.org.uk

European Mycological Association
Links to societies from all
over the world
http://www.euromould.org

First Nature
Large online resource
www.first-nature.com/fungi

NHBS
Everything for wildlife, science and
environment with specialist fungi
books
www.nhbs.com

Rogers Mushrooms
Online reference tool to learn about
and identify fungi
www.rogersmushrooms.com

The Mushroom Diary
UK blog and identification guide
www.mushroomdiary.co.uk

IMAGE CREDITS

Bloomsbury Publishing would like to thank the following for providing
photographs and for permission to produce copyright material. While every
effort has been made to trace and acknowledge all copyright holders, we would
like to apologise for any errors or omissions and invite readers to inform us so
that corrections can be made in any future editions of the book.

Key: l=left; r=right; FLPA = Frank Lane Picture Agency; SH = Shutterstock; NPL
= Nature Picture Library; Wiki = Wikimedia

4 Dave Pressland/FLPA; 16 Kletr/SH; 17 IMAGEBROKER/FLPA; 23 Richard Becker/FLPA; 24
Gianpiero Ferrari/FLPA; 31 Gianpiero Ferrari/FLPA; 32 Tony Hamblin/FLPA; 34 Jean-Yves
Grospas/Biosphoto/FLPA; 38 David Hosking/FLPA; 40 Gianpiero Ferrari/FLPA; 41
IMAGEBROKER/FLPA; 44 Robert Canis/FLPA; 49 Dave Pressland/FLPA; 50 Jean-Yves Grospas/
Biosphoto/FLPA; 57 Martin Fowler/SH; 60 Matthijs Wetterauw/SH; 62 PETRUK VIKTOR/SH;
A.S.Floro/SH; 66 Dariusz Majgier/SH; 67 Inna G/SH; 68 Eric Baccega/NPL; 72 Bob
Gibbons/FLPA; 76 Martin Fowler/SH; 77 Imagebroker/FLPA; 79 Yves Lanceau/Biosphoto/
FLPA; 82 Ingo Schulz/Imagebroker/FLPA; 85 Matthijs Wetterauw/SH; 88 Ron Wolf; 94 Kurt
Möbus/Imagebroker/FLPA; 98 Steen Drozd Lund/Biosphoto/FLPA; 100 Piotr Debowski/SH;
104 richsouthwales/SH; 105 Visuals Unlimited/NPL; 106 Philippe Clement/NPL; 108 Yves
Lanceau/Biosphoto/FLPA; 116 l Tony Hamblin/FLPA; 116 r Sterre Delemarre/Minden
Pictures/FLPA; 120 Francis Bossé/SH; 121 FLPA Dave Pressland/FLPA; 127 Martin Fowler/SH;
128 Stoelwinder/Biosphoto/FLPA; 129 Henrik Larsson/SH; 130 Philippe Clement/NPL; 131
Florian_am13/SH; 132 Ville Kangas/SH; 136 Andre Pascal/Biosphoto/FLPA; 138 Maryna
Pleshkun/SH; 140 ImageBroker/Imagebroker/FLPA; 143 Aleksander Bolbot/SH; 153 Sterre
Delemarre/Minden Pictures/FLPA; 154 Marcus Webb/FLPA; 161 Henk Verbiesen/Biosphoto/
FLPA; 162 l, r Dave Pressland/FLPA; 163 Steve McWilliam/SH; 164 l, r Dave Pressland/FLPA;
168 ImageBroker/Imagebroker/FLPA; 175 Roger Tidman/FLPA; 179 ImageBroker/
Imagebroker/FLPA; 179 Sterre Delemarre/Minden Pictures/FLPA; 180 Kurt Friedrich Möbus/
Imagebroker/FLPA; 187 Yves Lanceau/Biosphoto/FLPA

INDEX

INDEX

INDEX